Maryland Weather Service

The Climatology and Physical Features of Maryland

Maryland Weather Service

The Climatology and Physical Features of Maryland

ISBN/EAN: 9783337343002

Printed in Europe, USA, Canada, Australia, Japan

Cover: Foto ©berggeist007 / pixelio.de

More available books at **www.hansebooks.com**

The . . . Climatology

AND

Physical Features of Maryland

FIRST BIENNIAL REPORT

OF THE

MARYLAND STATE

WEATHER SERVICE

FOR THE YEARS 1892 AND 1893

U. S. DEPARTMENT OF AGRICULTURE, WEATHER BUREAU

THE

CLIMATOLOGY

AND

PHYSICAL FEATURES

OF

MARYLAND

FIRST BIENNIAL REPORT

OF THE

MARYLAND STATE WEATHER SERVICE

FOR THE YEARS 1892 AND 1893

BALTIMORE

1894

PRESS OF
THE FRIEDENWALD COMPANY
BALTIMORE, MD.

CONTENTS.

To His Excellency, FRANK BROWN, *Governor of the State
of Maryland:*

SIR:

I beg leave to present herewith, for transmission to the
Legislature, the First Biennial Report of the Maryland
State Weather Service for the years 1892 and 1893.

Respectfully submitted,

WILLIAM B. CLARK,

Director, Maryland State Weather Service.

JOHNS HOPKINS UNIVERSITY,
BALTIMORE, *December 30, 1893.*

BOARD OF GOVERNMENT.

LIST OF OBSERVERS.

Observers of U. S. Weather Bureau at Central Office.

C. P. Cronk, *Observer in Charge*, U. S. Weather Bureau, and Editor of Reports.

G. N. Wilson, *Observer*, U. S. Weather Bureau.

J. H. Donaldson, *Observer*, U. S. Weather Bureau.

A. T. Brewer, *Observer*, U. S. Weather Bureau, and Assistant Editor of Reports.

R. C. New, *Observer*, U. S. Weather Bureau, and Assistant Editor of Reports.

Meteorological Observers.

STATION.	COUNTY.	OBSERVER.
Agricultural College	Prince George's	W. H. Zimmerman, A. M
Annapolis	Anne Arundel	Walter Hay, M. D.
Bachman's Valley	Carroll	J. M. Myers
Baltimore		Central Office Observ's
Barron Creek Springs	Wicomico	Albert E. Acworth
Benedict	Charles	Thomas Berry
Boettcherville	Allegany	F. F. Brown
Cambridge	Dorchester	Calvert Orem
Charlotte Hall	St. Mary's	R. W. Silvester
		Maj. G. M. Thomas
		Prof. J. F. Coad
Chestertown	Kent	Hon. M. de K. Smith
Cumberland	Allegany	Howard Shriver
		E. T. Shriver
Darlington	Harford	A. F. Galbreath
Denton	Caroline	F. C. Ramsdell
Distributing Reservoir, D. C.		Col. Elliot
Dover, Del.	Kent	Jno. S. Jester

STATION.	COUNTY.	OBSERVER.
Easton	Talbot	S. P. Minnick
		G. W. Minnick
Edgemont	Washington	Chas. Feldman
Fallston	Harford	G. G. Curtiss, A. M.
Fenby	Carroll	Wm. Fenby
Frederick	Frederick	G. Ernest Bantz
Glyndon	Baltimore	A. W. Nyce
Great Falls	Montgomery	Col. Elliot
Hagerstown	Washington	Chas. Feldman
Jewell	Anne Arundel	Jos. Plummer
Kirkwood, Del.	New Castle	W. C. L. Carnagy
Leonardtown	St. Mary's	G. W. Joy
McDonogh	Baltimore	H. Pender
Milford, Del.	Kent	J. Y. Foulk
Millsboro, Del.	Sussex	Rev. L. W. Wells
Mt. St. Mary's	Frederick	J. A. Mitchell, A. M.
New Market	Frederick	H. H. Hopkins, M. D.
		Miss Marg't D. Hopkins
Oakland	Garrett	Jas. D. Hamill
		J. Lee McComas, M. D.
Penn's Grove. N. J.	Salem	Wm. T. Wilson
Princess Anne	Somerset	B. O. Bird
Receiving Reservoir,	D. C.	Col. Elliot
Salisbury	Wicomico	Lemuel Malone
		Randolph Humphreys
Seaford, Del.	Sussex	H. L. Wallace
Solomon's	Calvert	W. H. Marsh, M. D.
Sunny Side	Garrett	John G. Knauer
Taneytown	Carroll	C. W. Weaver, M. D.
		David H. Bowers
Upper Marlboro	Prince George's	Fred. Sasscer
		J. B. Perrie
Valley Lee	St. Mary's	Col. J. Edwin Coad
Washington, D. C.		S. W. Beall
Wilmington, Del.	New Castle	W. C. R. Colquhoun
Woodstock College	Baltimore	T. J. A. Freeman. S. J.
Birdsnest, Va.	Northampton	C. R. Moore
Cape Charles, Va.	Northampton	O. A. Browne
Norfolk, Va.	Norfolk	A. B. Crane
Warsaw, Va.	Richmond	C. H. Constable

Crop Correspondents.

STATION.	COUNTY.	CORRESPONDENT.
Aberdeen	Harford	E. E. Carsins
Adamstown	Frederick	R. C. Dutrow
Agricultural College	Prince George's	E. W. Doran
Annapolis	Anne Arundel	W. B. Finkbine
		D. S. Sprogle
Bagley	Harford	G. G. Curtiss, A.M.
Bark Hill	Carroll	W. H. Engler
Barron Creek Springs	Wicomico	A. E. Acworth
Beaver Creek	Washington	Joseph Witmer
Bellevue	Talbot	Forrest Scott
Benfield	Anne Arundel	B. F. Pumphrey
Berlin	Worcester	Thos. G. Hanley
Bittinger	Garrett	P. P. Lohr
Boettcherville	Allegany	F. F. Brown
Boonsboro	Washington	G. W. McBride
Bradshaw	Baltimore	B. F. Taylor
Bristol	Anne Arundel	E. O. Welch
Brooklyn	Anne Arundel	Wm. S. Crisp
Brookview	Dorchester	Geo. E. Lord
Bryantown	Charles	B. M. Edelen, Jr.
		H. A. Turner
Buckeystown	Frederick	R. C. Dutrow
		T. S. Bacon
		Jos. N. Chiswell
Cambridge	Dorchester	C. S. Jackson
Cape Charles, Va.	Northampton	O. A. Browne
Carrollton	Carroll	J. L. Albaugh
		D. E. Walsh
Catonsville	Baltimore	A. L. Crosby
Cavetown	Washington	G. R. Crowther
Chesapeake City	Cecil	J. W. Harriott
Clemsonville	Frederick	L. A. Bostion
Coleman	Kent	C. W. Harris
Colora	Cecil	Elwood Balderston
		Geo. Balderston
Contee	Prince George's	S. W. Beall
Cornersville	Dorchester	J. M. Beckwith
		G. D. Nutter

STATION.	COUNTY.	CORRESPONDENT.
Cumberland	Allegany	Howard Shriver
		M. C. Hendrickson
Dailsville	Dorchester	C. S. Jackson
Darlington	Harford	A. F. Galbreath
Davidsonville	Anne Arundel	P. H. Israel
Delight	Baltimore	Geo. Hempfling
Denton	Caroline	F. C. Ramsdell
Dickerson	Montgomery	L. B. Scholl
Drum Point	Calvert	Alex. de Barril
Easton	Talbot	S. P. Minnick
		G. W. Minnick
Edgemont	Washington	Chas. Feldman
Edgewood	Harford	Mrs. S. S. Russell
Elvaton	Anne Arundel	G. T. Sappington
Fallston	Harford	G. G. Curtiss, A.M.
Federalsburg	Caroline	G. F. Quidort
Fenby	Carroll	Wm. Fenby
Frederick	Frederick	Douglass Hargett
Grantsville	Garrett	J. S. Miller
Greensboro	Caroline	A. B. Roe
Hagerstown	Washington	D. W. Reichard
		Chas. F. Lehman
Hancock	Washington	W. F. Humbert
Harris Lot	Charles	Jas. R. Perry
Havre de Grace	Harford	J. Wm. Mitchell
Huyett	Washington	M. H. Huyett
Hyattsville	Prince George's	G. B. Pfeiffer, B.Sc.
Indian Springs	Washington	W. F. Humbert
Jewell	Anne Arundel	A. C. Wilson
Keedysville	Washington	J. A. Miller
Kirkwood, Del.	New Castle	J. F. Nelson
Landover	Prince George's	H. D. Metcalf
Leonardtown	St. Mary's	Geo. W. Joy
Linwood	Carroll	W. C. Rinehart
Lock 53	Washington	W. B. Seavolt
Mattawoman	Charles	W. C. Tippett
Mechanicstown	Frederick	J. J. Henshaw
Middletown	Frederick	G. C. Rhoderick. Jr.
Milford, Del.	Kent	J. Y. Foulk
Millington	Kent	John C. Turner
Mountain Lake Park	Garrett	Chas. J. Bunce
Mt. Pleasant, Del.	New Castle	A. S. Eliason

STATION.	COUNTY.	CORRESPONDENT.
Mt. St. Mary's	Frederick	J. A. Mitchell, A. M.
New Market	Frederick	H. H. Hopkins, M. D.
		Miss M. D. Hopkins
Patuxent	Anne Arundel	R. T. Donaldson
Pikesville	Baltimore	Chas. T. Cockey
Pleasant Valley	Carroll	N. H. Kester
Port Deposit	Cecil	E. Noyes, Jr.
Princess Anne	Somerset	B. O. Bird
Queenstown	Queen Anne's	Chas C. Willson
Rawlings	Allegany	R. C. Wilson
Rising Sun	Cecil	G. E. Fisher
		W. P. Reynolds
		W. T. B. Roberson
Roberts	Queen Anne's	Jas. T. Scott
Rutland	Anne Arundel	Benj. Watkins
Salisbury	Wicomico	Lemuel Malone
Singerly	Cecil	Jas. M. Naudain
Slidell	Montgomery	J. S. Carlin, M. D.
Sunny Side	Garrett	John G. Knauer
Swanton	Garrett	R. Beckman
		Chas. T. Sweet
Sykesville	Carroll	Chas. R. Favour
		E. M. Mellor
Taneytown	Carroll	C. W. Weaver, M. D.
		David H. Bowers
		P. B. Englar
Thurston	Frederick	W. J. Sumwalt
Trappe	Talbot	Percival Mulliken
Tunis Mills	Talbot	Henry Rieman
Twilley	Wicomico	G. C. Twilley
Upper Falls	Baltimore	H. A. Wroth
Walnut Landing	Dorchester	J. J. Bennett
Westminster	Carroll	E. C. Bixler
Westwood	Prince George's	G. B. Pfeiffer, B. Sc.
Woodsboro	Frederick	G. F. Smith

Weather Signal Displaymen.

STATION.	COUNTY.	DISPLAYMAN.
Annapolis	Anne Arundel	W. M. Abbott
Appleton	Cecil	W. C. Henderson
Baltimore		Central Office
Barron Creek Springs	Wicomico	L. A. Wilson
Bel Air	Harford	N. N. Nock
Bradshaw	Baltimore	B. F. Taylor
Bridgeville, Del.	Sussex	T. J. Gray
Buckeystown	Frederick	A. W. Nicodemus
Cambridge	Dorchester	Calvert Orem
Chestertown	Kent	J. S. Vandegrift
Darlington	Harford	A. F. Galbreath
Delaware City, Del.	New Castle	W. E. Reybold
Dickerson	Montgomery	W. H. Dickerson
Dover, Del.	Kent	Philip Burnet
Easton	Talbot	G. W. Minnick
* Felton, Del.	Kent	J. H. Hubbard
Frederick	Frederick	W. T. Delaplaine
Frostburg	Allegany	C. J. Conner
Glyndon	Baltimore	A. W. Nyce
		John J. Dyer
Grantsville	Garrett	T. H. Bittinger
Greensboro	Caroline	Plummer & Plummer
Hagerstown	Washington	R. J. Hamilton
Havre de Grace	Harford	W. S. McCombs
Hyattsville	Prince George's	E. B. Rowell
Lonaconing	Allegany	J. J. Robinson
Middletown	Frederick	G. C. Rhoderick, Jr.
Milford, Del.	Kent	J. Y. Foulk
Mt. St. Mary's	Frederick	Jos. H. Martin
Oakland	Garrett	J. M. Litzinger
		J. Lee McComas, M. D.
Odenton	Anne Arundel	E. B. Watts
Ridgely	Caroline	J. A. Sigler
Rising Sun	Cecil	E. A. Reynolds
Rockville	Montgomery	Emmit Dove

* Whistle Signals only.

STATION.	COUNTY.	DISPLAYMAN.
Salisbury	Wicomico	L. W. Gunby
Seaford, Del.	Sussex	Dr. Hugh Martin
		H. L. Wallace
Snow Hill	Worcester	Purnell & Vincent
* Sparrow's Point	Baltimore	Md. Steel Co.
St. Michael's	Talbot	E. M. Jefferson
Taneytown	Carroll	C. W. Weaver, M. D.
Westminster	Carroll	W. S. Myer & Bro.
Westover	Somerset	E. D. Long
Wilmington, Del.	New Castle	Wm. Lawton
Woodsboro	Frederick	G. F. Smith

*Whistle Signals only.

All who have officiated during the past two years have been included in the foregoing lists.

REPORT OF THE DIRECTOR.

The Maryland State Weather Service was organized May 1, 1891, under the joint auspices of the Johns Hopkins University, the Maryland Agricultural College, and the United States Weather Bureau.

The few scattered observers in Maryland and Delaware, who had hitherto reported to the Chief of the United States Weather Bureau, were authorized on that date to send their reports to the central office at the Johns Hopkins University. At the same time the Baltimore office of the United States Weather Bureau was moved to the University, as the future efficiency of the State Service was recognized to depend largely upon the closeness of co-operation with the National Service. Quarters were assigned in the Physical Laboratory, and the roof of that building has since been used for the exposure of instruments.

Two series of reports were at once established: First, a monthly Meteorological Report, which began with May and was continued until November; the second, a weekly Crop Bulletin, the first issue of which appeared on June 26, and was continued on every succeeding Saturday until September 25. These reports were sent widely throughout the State and elicited much favorable comment from the people and the press.

It was evident, from the start, that the results of the local service could not be made available to the people of the State unless they provided means for the publication and distribution of the information obtained. The institutions interested in the organization of the State Service were willing to prepare the data for publication, but they had no fund at their disposal for printing. To that end, a bill was introduced in the last Legislature, was passed by both houses and was signed by the Governor. It provides for the establishment of the Maryland State Weather Service, the commissioning of its officers by the Governor, and an appropriation to defray the expenses of printing. It reads as follows:

AN ACT TO ESTABLISH A STATE WEATHER SERVICE, AND TO MAKE AN APPROPRIATION THEREFOR.

SECTION 1. *Be it enacted by the General Asssembly of Maryland*, That there is hereby established a State Weather Service, which shall be under the control and management of the Johns Hopkins University, the Maryland Agricultural College and the United States Weather Bureau. The officers of said service shall be a Director, designated by the President of the Johns Hopkins University, a Secretary and Treasurer, designated by the President of the Maryland Agricultural College, and a Meteorologist in Charge, designated by the Chief of the United States Weather Bureau; they shall be commissioned by the Governor, and be duly qualified as officers of the State. The said officers shall constitute a Board of Government, under the direction of the institutions from which they are appointed, and shall receive no compensation for their services as such officers.

SEC. 2. The central station and office of said service shall be at the Johns Hopkins University. The Board of Government shall establish, if practicable, one or more voluntary meteorological stations in each county in the State, and supervise the same, co-operating with the Chief of the United States Weather Bureau for the suitable location of such stations, in order that the greatest usefulness may result to the State and National services. The said officers are authorized to print weekly and monthly reports of the results and operations of said services, and to distribute the same in such manner as they shall deem most serviceable to the people of the State.

SEC. 3. The sum of two thousand dollars * annually, or so much thereof as shall be necessary, is hereby appropriated out of any funds of the Treasury not otherwise appropriated, for the purpose of carrying out the provisions of this Act, to be paid to said officers, or to their order, by the Treasurer, upon the warrant of the Comptroller, and upon the vouchers of said officers; provided, however, that no part of said sum shall be paid for salaries for any officer or officers, but a reasonable compensation may be paid for printing and other necessary and proper expenses of said officers.

SEC. 4. The said officers shall report to the Legislature at its regular sessions their expenditures under the provisions of this Act, and such other information as said officers may deem desirable, or as the Legislature may require.

SEC. 5. *And be it further enacted*, That this Act shall take effect from the date of its passage. *Approved April 7*, 1892.

The Maryland State Weather Service is similar in its organization and methods to like services in other States. The personnel consists of voluntary observers, who have been selected at favorable points throughout the State, and of a corps of permanent observers who have been assigned from the National Bureau to take charge of the work at the central office. To obtain the results of such detailed study of the climatic features of the State, the National Bureau supplies instruments, forms and stationery to all the stations of the service, and all the correspondence and reports are mailed under the frank of the Agricultural Department. As a result of such liberal encouragement on the part of the general government, it remains only for the State to bear the expense of publishing these results in bulletins and reports for general distribution.

The stations connected with the service are of three classes: 1st, those which report meteorological facts; 2nd,

---

* The Governor signed the bill on the condition that only $1000 of this appropriation should be used annually. The restriction was later withdrawn, so that the full amount was available the second year.

those which send crop notices; 3rd, those which display
signals. In a few instances the same person fulfills the
duties of all three offices. Connected with the first class
there have been 58 observers during the past two years, with
the second class 111 correspondents, and with the third class
43 displaymen, making 212 persons connected with the
Maryland State Weather Service outside of the officers
at the central office. To all of these the fullest praise is
due for the important aid they have rendered in the preparation of the weekly and monthly forms and publications
since the organization of the service. Nearly every section
of the State is represented, but it is hoped that the number
of observers will steadily increase.

The publications of each State Service are chiefly
devoted to a discussion of the climate of the territory
covered by its observations, the effect of the meteorological conditions upon the products of the soil, and the
special advantages to be enjoyed by the inhabitants. On
account of the varied climate of Maryland, the differences
in its soil formation, and its extensive coast-line, the agricultural and commercial interests of the State are many
and important, and a well-equipped State Weather Service
may be of great value in bringing to the attention of the
public the special advantages which the State possesses in
these directions. It has been the aim of the Maryland
State Weather Service during the past two years to dissem-
inate widely such information by means of its various
publications.

The sum appropriated by the State government has
enabled the printing of monthly Meteorological Reports,
extending throughout the year, and of weekly Crop Bulletins, extending throughout the growing and harvesting

seasons. That these publications are appreciated by the people of the State is evidenced from the hearty support they have given in rendering information in regard to the weather and the crops, and by the general and complete publication of reports by the newspapers of the State. Two thousand copies of each of the monthly Meteorological Reports have been regularly published, while the edition was increased to six thousand during the period of the World's Fair, and a series of special articles written, setting forth the advantages of the State from a climatic standpoint. During the first year about five hundred copies of the weekly Crop Bulletin were printed, but this number was gradually increased to twelve hundred the second year, as the demand for them grew.

In addition to the regular publications, one hundred sets of a series of ten large Climatic Charts with explanatory text have been prepared, which represent the seasonal and annual temperature and rainfall of the State. It would be advisable to publish a much larger edition of these maps for distribution to public schools and other educational institutions, and for such other uses as would render them available to the public generally. The presentation in this graphic way of the typical features of Maryland climate will be much more readily comprehended than the tabulated statements.

The present Biennial Report is intended to be a general review of the climate of Maryland, so far as conclusions can be drawn from the data thus far attainable. It is illustrated by maps, diagrams and tables which will show at a glance the leading climatic features of the State. The data for the general chapters upon the Physical Features of Maryland were largely adapted from the articles written

by Professors Wm. Hand Browne, Geo. H. Williams, Milton
Whitney and William B. Clark for the World's Fair Book
of Maryland, and the Monthly Meteorological Reports.
The later portions of the Report containing the sum-
maries of the weather and the crops and, also, the tables
for the past two years, were prepared under the super-
vision of Dr. C. P. Cronk, the Meteorologist in charge.

These various regular and special reports have been
sent widely throughout Maryland, into other States and to
foreign countries. They have been most favorably received
and have added largely to the spread of information con-
cerning the State of Maryland.

This information would reach more widely those
elements in foreign countries from whom immigration is
desired, if certain of the publications were translated into
German and other languages. Letters, bearing upon this
point, have been obtained from consuls in European countries, showing the important use to which the reports could
be put if they were in the language of the people.

Much fuller knowledge could be given of the physical
geography of the State if colored maps, showing the areas
of highland and lowland and the leading drainage basins,
could be prepared to accompany the climatic charts. The
data are at hand for the representation of these features
and the maps could be readily prepared.

Great advance has been made during the past two
years in the knowledge of the physical features of Maryland, and with the steady growth of interest among all
classes of people and the gradual increase in the number
of stations, much greater progress may be looked for in
the future.

WILLIAM B. CLARK.

THE PHYSICAL FEATURES OF MARYLAND.

THE TOPOGRAPHY.

Maryland is situated between the parallels of 37°53′ and 39°44′ north latitude, and the meridians of 75°04′ and 79°33′+ west longitude, the exact western boundary being yet undetermined. Its boundaries are: Mason and Dixon's line, separating it from Pennsylvania on the north; the State of Delaware and the Atlantic Ocean on the east; on the south a line drawn westward from the ocean to the western bank of the Potomac river, thence following the western bank of that river to its source; and on the west, a line drawn due north from this source to Mason and Dixon's line. The gross area of the State is 12,210 square miles, of which 9860 square miles are land surface; the included portion of Chesapeake Bay, 1203 square miles; Assateague Bay on the Atlantic coast, 93 square miles; with 1054 square miles of smaller estuaries and rivers.

The eastern borders of the State are deeply indented by bays and estuaries, and perhaps nowhere else in the world is there a coast-line proportionately so extensive, or any country offering such facilities for water transportation as tide-water Maryland.

Maryland is divided into twenty-three counties, of which Garrett, Allegany, and Washington form the mountainous region known as Western Maryland; Frederick, Carroll, Baltimore, Harford, Cecil, Howard, and Montgomery the Piedmont area, which is also referred to under the name of Northern-Central Maryland; Anne Arundel, Prince George's, Calvert, Charles, and St. Mary's, commonly called Southern Maryland; and Kent, Queen Anne's, Talbot, Caroline, Dorchester, Wicomico, Somerset, and Worcester, known as Eastern Maryland. Of these twenty-three counties there are but seven which do not lie upon navigable waters.

The three leading topographic regions of the eastern portion of the United States, viz: the Coastal Plain, the Piedmont Plateau, and the Appalachian Region, are all typically represented within the limits of the State of Maryland.

The Coastal Plain.—The Coastal Plain forms the eastern portion of the State, and comprises the area between the Atlantic Ocean and a line passing N. E. to S. W. from Wilmington to Washington through Baltimore. This region includes very nearly 5000 square miles, or, approximately, one-half the area of the State. It is about 100 miles broad in its widest part.

The Coastal Plain is characterized by broad, level-topped stretches of country, which extend, with gradually increasing elevations, from the coastal border, where the land is but slightly raised above sea-level, to its western edge, where heights of 300 feet and more are found. As the region is cut quite to the border of the Piedmont Plateau with tidal estuaries, the topography becomes more and more pronounced in passing inland from the coast.

The Coastal Plain, in Maryland, may be divided into a lower eastern and a higher western division, separated by the Chesapeake Bay. The former has been referred to under the name of Eastern Maryland (or Eastern Shore), while the latter is called Southern Maryland.

The *eastern division*, or Eastern Maryland, except in the extreme north, does not reach at any point 100 feet in elevation, while most of the country is below 25 feet in height. Both on the Atlantic coast and the shore of the Chesapeake it is deeply indented by bays and estuaries.

The drainage of the region is simple, the streams flowing from the watershed directly to the Atlantic Ocean and Delaware Bay upon the east, and to the Chesapeake Bay upon the west, while the position of the watershed, along the extreme eastern edge of the area, is a very striking feature. Among the more important rivers which reach the Chesapeake Bay are the Pocomoke, Nanticoke, Choptank, and Chester, all of which have their headwaters within the State of Delaware, and flow in a general southwest direction in sinuous channels.

The *western division*, or Southern Maryland, stands in striking contrast to the eastern division, since it frequently exceeds 100 feet in height, even along its eastern margin. In lower St. Mary's county the land reaches an elevation of 100 feet not far from the Bay shore, which is gradually increased until, near the border of Charles county, the region slightly exceeds 180 feet. In southern Calvert county an elevation of 140 feet is found to the west of Cove Point, and this gradually increases to the northward, until near the southern boundary of Anne Arundel county the land rises above 180 feet. Farther to the northwest, in Charles, Prince George's, and Anne Arundel counties, the land increases gradually in height, reaching at several points above 250 feet.

The western division is traversed by several rivers which flow from the Piedmont Plateau. Among the more important are the Potomac, Patuxent, Patapsco, Gunpowder, and Susquehanna.

The local drainage of the western division is similar to that hitherto described for the eastern, in that the streams throughout Southern Maryland flow chiefly to the westward.

The Piedmont Plateau.—The Piedmont Plateau borders the Coastal Plain upon the west, and extends to the base of the Catoctin Mountain. It includes, approximately, 2500 square miles, or about one-fourth of the area of the State. It is nearly 40 miles in width in the southern portion of the region, but gradually broadens toward the north, until it reaches 65 miles. It includes all, or the greater part of Montgomery, Howard, Baltimore, Harford, Cecil, Carroll, and Frederick counties. The country is broken by low, undulating hills, which gradually increase in elevation to the westward.

The Piedmont Plateau, in Maryland, is divided very nearly in its central portion by an area of highland known as Parr's Ridge, into an eastern and a western district. In the character of the rocks these divisions afford sharp distinctions, which are not without their effect upon the relief of the land.

The *eastern division* of the Piedmont Plateau has, on account of its crystalline rocks and their complicated structure, a diversified topography. Along the eastern margin the land attains, at several points, heights exceeding 400 feet, reaching at Catonsville 525 feet above sea-level. To the west the country gradually increases in elevation, until it culminates in Parr's Ridge, which exceeds 850 feet in Carroll county.

The drainage of the eastern district is to the east and southeast. On its northern and southern boundaries it is traversed by the Susquehanna and Potomac rivers, which have their sources without the area, while the smaller streams which lie between them either drain directly to the Chesapeake Bay or into the two main rivers. Among the larger of the intermediate streams are the Patuxent, Patapsco, and Gunpowder rivers, whose headwaters are situated upon Parr's Ridge. The Patapsco, especially, flows in a deep rocky gorge until it reaches the Relay, where it debouches into the Coastal Plain. All these streams have rapid currents as far as the eastern border of the Piedmont Plateau, and even in the case of the largest rivers are not navigable within that district.

The *western division* of the Piedmont Plateau extends from Parr's Ridge to Catoctin Mountain. Along its western side is the broad limestone valley in which Frederick is situated, and through which flows the Monocacy river from north to south, entering the Potomac river at the boundary line between Montgomery and Frederick counties. The valley near Frederick has an elevation of 250 feet above tide, which changes slowly to the eastward toward Parr's Ridge, and very rapidly to the westward toward Catoctin Mountain. Situated on the eastern side of the valley, just above the mouth of the Monocacy river, and breaking the regularity of this surface outline, is Sugar Loaf Mountain, which rises rapidly to a height of 1250 feet.

With the exception of a few streams which flow into the Potomac directly, the entire drainage of the western district is accomplished by the Monocacy river and its

numerous tributaries, which flow in nearly parallel west
and east courses, from Parr's Ridge and the Catoctin
Mountain.

The Appalachian Region.—The Appalachian Region
forms the western portion of Maryland, bordering the
Piedmont Plateau. It comprises about 2000 square miles
or, approximately, one-fifth the area of the State. It includes the western portion of Frederick, and all of Washington, Allegany, and Garrett counties. It consists of a
series of parallel mountain ranges with deep valleys, which
are cut nearly at right angles by the Potomac river. Many
of the ranges exceed 2000 feet, while some reach 3000 feet
and more, in the western portion of the mountainous area.

The Appalachian Region is divided into three distinct
districts, an eastern (Blue Ridge and Great Valley), a central (Appalachian Mountains proper), and a western (Allegany Mountains), which are separated from one another
upon clearly defined structural differences.

The *eastern division* comprises the area between the
Catoctin and the North Mountains, and has a width of
about 25 miles from east to west. Along the eastern border
of this region the Catoctin Mountain extends from north
to south, reaching the Potomac river at Point of Rocks.
It attains an altitude of 1800 feet. Succeeding this range
upon the west is the Middletown Valley, with an elevation
of 500 feet at Middletown. Running through its center
from north to south is the Catoctin Creek, which receives
the drainage from the western flanks of the Catoctin Mountain and the eastern slope of the Blue Ridge. The Blue
Ridge Mountains are a continuation of the South Moun-

tains of Pennsylvania, and extend as a sharply defined range from the northern boundary of the State to the Potomac river, which they reach at Weverton. Their crest forms the boundary between Frederick and Washington counties. The Blue Ridge reaches an elevation of about 2400 feet at Quirauk.

Occupying the greater part of this eastern district, and reaching to its western border, is the Hagerstown Valley, a portion of the Great Valley of the Appalachian Region hitherto described. It reaches an altitude of about 500 feet at Hagerstown, but gradually becomes lower toward the south in the vicinity of the Potomac river. The Antietam river and its tributaries occupy the eastern side of the valley, and the Conococheague river and its tributaries the western.

The *central division*, which comprises the Appalachian Mountains proper, extends from the North Mountain upon the east to Will's Mountain near Cumberland upon the west. Among the more important of the ranges in Maryland, west of North Mountain, are Tonoloway Hill, Sideling Hill, Town Hill, Green Ridge, Warrior Ridge, and Martin's Ridge, the two latter reaching 2000 feet and upwards in elevation. They are arranged in groups of three parallel and closely adjoining ridges on the east and west, with more distant ranges in the middle of the district.

The drainage is altogether to the southward into the Potomac. The deeper valleys in the eastern portion of the region have an elevation of about 500 feet in the vicinity of the Potomac, but they gradually become higher toward the west.

The *western division* occupies the extreme western portion of Maryland, and includes the Allegany Mountains
in its eastern half. They gradually merge into a high
plateau, with gently undulating mountains rising from the
surface, which continue beyond the western borders of the
State. The leading ranges of this district are Dan's Mountain, Savage Mountain, Meadow Mountain, Negro Mountain, Winding Ridge, and Laurel Hill. Heights of 3000 feet
and more are reached in Savage and Negro Mountains.

The partially adjusted streams give much variety to
the topography. They flow in part to the southward into
the Potomac, but in Garrett county the greater number
drain to the northward through the Youghiogeny river into
the Monongahela.

This separation of the drainage has particular interest,
since it marks the watershed between the streams which
flow into the Potomac and thus reach the sea by the
eastern slope of the Appalachian Mountains, and those
which flow to the Gulf by way of the Ohio and Mississippi
rivers.

THE GEOLOGY.

The State of Maryland is so situated as to display, in spite of its comparatively small size, a remarkably perfect sequence of all the geological formations. The most ancient rocks which go to make up the earth's crust, as well as those still in the process of deposition, are here to be found, while between these wide limits there is hardly an important geological epoch which is not represented. It is doubtful whether another State of the Union contains a fuller history of the earth's past.

The Piedmont Plateau.—The rocks composing the Piedmont Plateau are divisible into two distinct classes. In the eastern portion they are completely crystalline, and, whatever was their origin, they now retain no certain evidence of clastic structure. They disappear beneath the overlying deposits of unconsolidated sand, gravel and clay which compose the Coastal Plain.

In the western portion of the Piedmont Plateau the rocks are semi-crystalline, and, while they have been subjected to a certain amount of metamorphism and alteration, they still plainly show that they were once sediments of an ordinary type. They are principally confined to the western half of the Plateau region.

Eastern Division.—The various rock-formations composing the eastern division of the Piedmont Plateau cross Maryland from the southeast corner of Pennsylvania

and the north end of Delaware in a general southwest
direction.

They are petrographically divisible into six distinct
types. Three of these are of undoubtedly eruptive
origin, and may be designated, according to their chemical and mineralogical composition, as *gabbro, peridotite*
or *pyroxenite*, and *granite*. The three remaining types—
gneiss, marble, and *quartz-schist*—are completely crystalline, and therefore exhibit no certain traces of clastic
structure.

The prevailing rock is the gneiss. It enters the State
from the north in a very wide band, completely surrounding the Delta Peach Bottom slate area, but its breadth
rapidly contracts toward the Potomac. The remarkably
irregular forms of the marble areas, which are intercalated
in the gneiss complex, show how intricate the stratigraphy
of the latter really is.

The gneiss is sometimes quite constant in character
for considerable distances, but more usually it consists of
a succession of differently constituted layers.

The marbles of the eastern plateau region differ from
all the other marbles and limestones of Maryland in being
much more coarsely and perfectly crystalline. They have
lost all evidence of an originally clastic structure through
recrystallization.

These highly crystalline marbles are of the same age
as the gneisses, and are infolded with them. In consequence of their greater solubility, they have been easily
removed, and now occupy depressions like the Green
Spring, Worthington, Mine Bank, Dulany's and other
valleys which are sharply bounded by the surrounding
ridges of gneiss.

The three types of eruptive rocks, which are distinguished on the geological map in the eastern plateau region, have all broken through and have more or less modified the gneisses, and are hence younger than these rocks.

The oldest, as well as the most extensive of the three eruptive rocks which so abundantly intrude the gneiss complex is the gabbro. Of this there are three main areas in Maryland—the Stony forest area of Harford and Cecil counties; the great belt or sheet which extends from north of Conowingo, on the Susquehanna river, in a south-southwest direction to Baltimore City, and the irregular intrusive area which is mainly developed to the west of Baltimore, but extends thence as far south as Laurel.

The action of pressure, which has caused the recrystallization of the gneiss and marble, is also very marked in the gabbro. It has caused its iron constituent, pyroxene, to change to another green mineral called hornblende. This has, in some cases, left the rock as massive as at first, and in other cases it has rendered it schistose. This resulting rock is called *gabbro-diorite*.

The next eruptive rocks in point of age are the basic magnesian silicates, peridotite or pyroxenite, and their alteration products, serpentine and steatite. These are intimately associated with the gabbros, but occur most abundantly toward the western edge of the crystalline region. They do not occur in as large masses as the other eruptive rocks, but occupy numerous small areas like the " Bare Hills," "Soldier's Delight," and many others which it is unnecessary to enumerate. In Montgomery county the serpentine is also developed in the semi-crystalline schists of the western plateau district.

The youngest intrusive rocks which break through the gneiss are the granites. They form large masses at Port Deposit and Havre de Grace, on the Susquehanna river; also near Joppa and to the north of Towson; at Woodstock and Sykesville; at and south of Ellicott City, and at several localities near Washington. The granites are so like the surrounding gneisses in chemical as well as in mineralogical composition that where they have been greatly foliated through dynamic action, it becomes a matter of no small difficulty to distinguish them.

The gneisses of the Baltimore region are penetrated with a great abundance of dykes, veins and "eyes" of the coarse-grained granite known as *pegmatite*. The other crystalline rocks of the region, although to a less extent, contain the same material.

The only other type of eruptive rock occurring in the eastern division of the Piedmont Plateau in Maryland is a long dyke of unaltered diabase, which preserves all the features of the normal triassic diabase occurring further west, and which is therefore referred to this age.

Western Division.—The western slope of Parr's Ridge, as far as the Monocacy river, is composed of little crystalline or semi-crystalline rocks of sedimentary origin. These rocks are almost unaltered along their western margin, where they present the same characters as the sandstones, slates and limestones of the Blue Ridge and Frederick Valley, where their age has been determined by fossils. Along their eastern margin dynamic action has been at a maximum, as is shown by the greatly contorted condition of the schists and by the abundant development of new minerals within them. The slates have become roofing-slates or chlorite and hydromica (sericite) schists. The limestones have become compact, hard, fine-grained marbles. The geological position of these rocks has not yet been positively proved by fossils, and they are, therefore, designated on the geological map by a different color. Nevertheless, there is no reasonable doubt that they are Cambrian sandstones, Trenton limestones, and Hudson River shales in a more or less completely metamorphosed form.

The deposit of red sandstone and shale, together with the limestone breccia ("Potomac marble" or "calico rock"), which covers a portion of the Frederick valley, lying along the east base of Catoctin Mountain, and spreading over a much wider area toward the north, represents an era geologically much more recent than any we have thus far considered. The strip of these red rocks, which here crosses Maryland, represents a formation laid down in estuaries after the Appalachians had been elevated. Through all Paleozoic time, the sediments of which these mountains are formed were accumulating. At the end of this time they were folded into a lofty range, at whose base the sandstone was deposited. The fossils and stratigraphical relations of this sandstone, known as the "Newark formation," show its geological position to be at the end of the first age of Mesozoic time or the Triassic. Its features are remarkably persistent along the whole Atlantic border.

The Appalachian Region.—The section made by Maryland across the Appalachian system, between the Frederick valley and the western line of Garrett county, presents an almost complete and unbroken sequence of the sediments which accumulated during the entire duration of Paleozoic time. The oldest of these sediments are toward the east and the youngest toward the west, although the more or less abrupt folds into which they were thrown when they were raised into a mountain-chain have since been so cut off by erosion as to show a repeated succession of strata.

The beds of sediments (limestones, sandstones and shales) which form the Appalachians were deposited in a wide, long trough which once extended over the region now occupied by these mountains. This trough was undergoing gradual depression through the whole of Paleozoic time, until about 40,000 feet of conformable beds had accumulated in it, mainly from the debris of a continental mass lying to the east. This vast accumulation was, at the end of the coal age, so compressed as to be forced up into a lofty range of mountains. The present Appalachians are merely the remnants of this range worn down by waste through many successive periods.

The Blue Ridge and Great Valley.—This division is composed of the oldest Paleozoic strata, embracing the Cambrian and Lower Silurian or Ordivician deposits.

The oldest, or Cambrian, consists mainly of hard sandstone, which forms the summits of the Blue and Catoctin mountains.

The Cambrian grades upward through its top member into the Lower Silurian limestone (Trenton-Chazy). The sandstones are cut by a number of great faults along which they have been thrust forcibly toward the west, and thus often overlie rocks which are in reality younger than themselves.

From all the central portion of the Blue Ridge in Maryland the sandstones and shales have been removed, and the older crystalline rocks upon which they rest are revealed. These consist in part of ancient volcanic rocks and in part of granites. The volcanic rocks are of two types, one corresponding to the acid lavas (rhyolite) and the other to the basic (basalt) of recent volcanic regions.

The broad and fertile Cumberland or Hagerstown valley is mainly composed of the blue Silurian limestone, which also underlies the Frederick valley, together with long, narrow areas of infolded shale which belong to the next younger or Hudson River division. None of the overlying Triassic red sandstone is, however, found here.

The Appalachians Proper.—This province of the Appalachian system includes the central portion of the mountains which lies between the western edge of the Great Valley (North Mountain) and Dan's Mountain west of Cumberland.

The eastern division of this region, the North Mountain belt, is about fifteen miles wide, and extends from the western edge of the Hagerstown Valley to the western slope of Tonoloway Hill. The strata are so closely folded that almost all the Silurian and Devonian beds are brought repeatedly to the surface.

The middle belt of the Appalachian province proper embraces the three ridges known as Sideling and Town Hills and Polish Mountain, with their intervening valleys. The strata in this belt are almost altogether Devonian, and their comparatively gentle undulations offer a contrast to the sharper folds on both the east and west sides.

The western belt of the Appalachian province proper in Maryland is closely folded like that on the east. It occupies the area between Flintstone and Dan's Mountain, and consists of three parallel and sharply anticlinal ridges—Warrior's, Martin's, and Will's mountains—whose axes consist in each case of hard Medina sandstone, while the younger strata from Clinton to Chemuing encircle their southern ends and fill the intervening valleys. Each of these three folds is steepest on its western side, as is the rule of Appalachian structure.

The Allegany Mountains.—The whole of Maryland west of Will's Creek belongs to the great plateau of the Alleganies, which gently undulates across West Virginia and Pennsylvania into the plains of Ohio. The rocks are the youngest of the Paleozoic series—Upper Devonian and Carboniferous. They are sandstones, conglomerates and shales, with very little limestone, and carry near their top the seams of coal. These strata in Western Maryland are bent into very gentle folds, and as the middle beds (Pottsville) are the hardest, these have most successfully resisted the forces of erosion, and stand out in bold ridges. Between these are alternately flat anticlinal valleys of older rocks and synclinal valleys of younger ones.

The Coastal Plain.—The area of lowland which borders the Piedmont Plateau upon the east passes with constantly decreasing elevation seaward, and has been already described under the name of the Coastal Plain. It is made up of geological formations of younger date than those found in the central and western portions of the State. These later sediments form a series of thin sheets, which are inclined slightly to the eastward, so that successively later formations are encountered in passing from the interior of the country toward the coast.

The Lower Cretaceous (Potomac).—The Lower Cretaceous or Potomac formation directly overlies the crystalline rocks of the Piedmont Plateau, and is to a considerable extent formed of debris from them.

The deposits consist chiefly of sands and clays, with gravels at certain points where the shore accumulations are still preserved. The sands and clays alternate, and show both a vertical and horizontol gradation into one another. The sand layers are seldom widely extended, being generally found as lenticular masses, which rapidly diminish in thickness from their centers. Highly colored and variegated clays (iron ore clays of Tyson) abound in the upper portion of the formation, and have yielded large amounts of nodular carbonate of iron.

The clays have also great value for pottery and brick-making, and a large part of the local materials so used comes from this formation.

Beds of sand are found underlying the clays at Federal Hill and other points, while even larger deposits occur in the upper portion of the series on the Patuxent and Severn rivers.

The width of outcrop of the Lower Cretaceous is about 15 miles in the center of the State. It extends from northeast to southwest across the State, the main body of the formation being found in Cecil, Harford, Baltimore, Anne Arundel, and Prince George's counties. Upon the eastern shore of the Chesapeake it reaches into Kent county, and along the Potomac into Charles county.

The fossils found in the deposits, although not as numerous or distinctive as might be desired, yet indicate beyond doubt the Cretaceous age of the formation. They consist chiefly of the bones of dinosaurian reptiles and leaf impressions.

The Upper Cretaceous (Severn).—The Upper Cretaceous or Severn formation rests unconformably upon the Lower Cretaceous.

The materials out of which the deposits of the Upper Cretaceous are formed consist of fine sands and clays, clearly stratified, and in the case of the clays often laminated. The clays and sandy clays are generally dark, often black in color. They are highly micaceous, indicating the crystalline rocks of the Piedmont Plateau as the source of the materials. The highly homogeneous and persistent character of the beds shows that deposition went on under similar and quiet conditions throughout a wide region of the sea-floor. In this respect the deposits of the Upper Cretaceous stand in marked contrast to those of the Lower Cretaceous, where every evidence of mechanical disturbance is present.

The Upper Cretaceous extends as a narrow band across Cecil and Kent counties, on the Eastern Shore, numerous fossils having been found at the head of Bohemia Creek and on the banks of Sassafras river. On the Western Shore it is found in Anne Arundel and Prince George's counties, characteristic fossils having been obtained at many points.

The Eocene (Pamunkey).—The third of the geological formations found represented in the Coastal Plain of Maryland is the Eocene or Pamunkey formation. It extends as a nearly unbroken belt from the Delaware line to the Potomac river, and is found in Cecil, Kent, Queen Anne's, Anne Arundel, Prince George's, and Charles counties. The remarkable green-sand and numerous fossils early attracted the attention of geologists. The breadth of out-crop upon the eastern shore of the Chesapeake is scarcely five miles at the head of the Sassafras river, but gradually expands toward the southwest until upon the western shore it is in places more than twenty-five miles wide.

The lithological character of the rocks is remarkably persistent. The typical deposit is a green-sand marl, which may, however, by chemical changes, lose its characteristic green color, and by the deposition of a greater or less amount of hydrous iron oxide be found as an incoherent red sand or firm red or brown sandstone. The green-sand type is chiefly confined to the southwestern portion of the area in Charles and Prince George's counties, where the deposits overlying the Eocene attain their greatest thickness. In Anne Arundel county and on the eastern shore of the Chesapeake the Eocene is less deeply buried and the strata are more thoroughly weathered, affording greenish gray or red sands, and at times bands of firm sandstone.

Green-sand has considerable economic importance as a fertilizer, but has never been employed to any extent in Maryland.

The Miocene (Chesapeake).—Occupying the region to the southeast of the Eocene, and extending across the State from northeast to southwest, is the Miocene or Chesapeake formation. The deposits of the Miocene lie unconformably upon those of the Eocene, and overlap them

along their western border, where they often rest upon
the Cretaceous.

The Miocene consists of sands, clays, marls and diatomaceous beds. The latter, composed chiefly of the shells
of microscopic plant forms called diatoms, are in the main
confined to the lower portion of the formation, and afford
fine sections at Pope's Creek, on the Potomac, at the mouth
of Lyon's Creek, a tributary of the Patuxent, and at Her-
ring Bay, on the western shore of the Chesapeake. At
these points the light-colored bluffs are very striking objects
in the landscape. The pure diatomaceous earth reaches a
thickness of about 30 feet, although the remains of diatoms
are found scattered in greater or less numbers throughout
much of the overlying strata. The deposits of diatomaceous earth have considerable economic value, and several
companies have been formed to work them at Pope's Creek
and on the Patuxent.

The greater portion of the Miocene is composed of
sands and clays of various colors, mingled with which are
frequently vast numbers of shells of calcareous organisms.
Extensive beds of marl thus formed are found outcropping
at many points. Sometimes the shelly materials form
so large a portion of the deposit as to produce almost
pure calcareous layers, which, in a partially comminuted
state, may become cemented into hard limestone ledges.
The deposits are at times very carbonaceous and dark in
color.

The Pliocene (Lafayette).—Widely covering the deposits of the Coastal Plain hitherto described is a formation
composed of gravel, sands and clays, which thus far has
afforded no fossils in the State of Maryland to indicate its

geologic age. From the fact that the deposits rest unconformably upon the underlying Miocene, and are in turn unconformably overlain by the Pleistocene, they have been considered to represent the Pliocene.

Within the State of Maryland the strata cover the higher levels upon the western shore of the Chesapeake, but have not been hitherto described from the eastern counties. Toward the ancient coast line, bordering the Piedmont Plateau, are deposits of coarse gravel, through which is scattered a light-colored sandy loam, the whole at times cemented by hydrous iron oxide into a more or less compact conglomerate.

The deposits are very irregularly stratified, and often change rapidly within narrow limits. The eastward extension of the formation shows a lessening of the coarser elements and a larger admixture of loam. The thickness of the deposits is estimated to reach about 25 feet in Maryland, which becomes considerably increased to the southward.

The Pleistocene (Columbia).—Superficially overlying the other deposits of the Coastal Plain is the Pleistocene, or Columbia formation, which, with marked variations in thickness, composition and structure, extends from the glacial accumulations of central New Jersey southward through the south Atlantic and Gulf States to the Mexican boundary.

Three distinct phases have been recognized, the *fluvial phase*, the *interfluvial phase*, and the *low-level phase*.

The *fluvial phase* is found in its fullest development along the leading waterways and their larger tributaries. It consists of a lower horizon of coarse materials, pebbles and boulders, passing upward into a brownish loam that at times becomes orange-yellow in color.

The *interfluvial phase* is found typically represented in the country which lies between the waterways, and is characterized by materials of local origin and produced largely by wave-action.

The *low-level phase* is developed throughout the area that is removed from the Pleistocene coast-line, and excellent sections of it are found along the shores of the Chesapeake Bay, where it buries to sea-level much of the region in the more southern of the Eastern Shore counties. The deposits consist of sands, clays and loams, which are often clearly stratified, while their fossils show the distinctly marine conditions surrounding their origin.

Economic Products.—The minerals of economic value which are known to exist within Maryland's territory may be divided into three classes:

1st. Those which are now produced with profit, or which are susceptible of future development. In this class may be enumerated coal, iron ore, gold, building stone (granite rocks, marble, sandstone, slate and, possibly, serpentine), decorative stone (marble, serpentine and porphyry), limestone (for burning and flux), hydraulic cement, clay (brick and potter's clay, fire clay), sand (building and molding sand), porcelain materials (flint, feldspar, kaolin), soapstone, mineral water.

2d. Mineral products, formerly produced in Maryland, but which are not at the present time actively worked. As such may be mentioned—copper ore, chrome ore, ochre (mineral paint), diatomaceous earth (infusorial earth, tripoli), magnesium carbonate, asbestos (chrysotile).

3d. Minerals known to occur in Maryland, but not in quantities sufficient to warrant their production. To this class belong—lead ore (galena), zinc ore (zinc blende), mica, amber, plumbago (graphite, black-lead), manganese, antimony.

In the crystalline rocks of the Piedmont Plateau we find the most varied, if not the most valuable list. Here occur the most important building stones: the slates of Delta and Ijamsville; the granites of Port Deposit, Woodstock and Guilford; the gneiss of Baltimore; the marble of Cockeysville and Texas; the sandstone of Deer Creek; and the serpentine of Broad Creek and Bare Hills. In these oldest rocks occur also all the ores of gold, copper, chrome, lead and zinc. Much of the best iron ore also belongs here, while all the flint, feldspar, kaolin and mica in the State must be sought for in this horizon. These older or pre-paleozoic rocks again appear in the center of the Blue Ridge, where they make the Middletown valley, and here they yield traces of copper, antimony and iron, while some of the red porphyries occurring a little farther north would appear to be well worthy of the attention of architects as decorative stones.

The long sequence of paleozoic strata which form the Appalachian Region furnishes much good sandstone and limestone, two horizons of valuable cement rock, and at its top it carries what is now left by man and the eroding agencies of nature of the wonderful Cumberland coal basin and its 14-foot vein of solid coal. This same basin contains also deposits of fire-clay and iron.

As we trace the sequence of formations through the more recently formed portions of the State, we find them not devoid of mineral deposits of economic value. The variegated limestone breccia, known as "Potomac marble," and the best brown sandstone for building purposes found in Maryland, both belong to the oldest of post-paleozoic strata—the triassic belt of the Frederick valley and southern Montgomery county. The series of still unconsolidated beds which represents the lapse of time from the Lower Cretaceous period to the present, and which composes the Coastal Plain of eastern and southern Maryland, besides furnishing valuable lands for various agricultural interests, contains our principal supply of brick, potter's and fire-clay; of sand, marl and diatomaceous earth; and much of our best iron ore.

THE SOILS.

The great number of geological formations and the complex topography of the State give rise to a great many different types of soil, each being especially adapted to certain classes of plants which give a very marked local distribution of crops in Maryland. Soils are composed of minute fragments of rocks and minerals which contain the mineral foods (potash, phosphoric acid, lime, iron, etc.), necessary for the life and growth of plants. Samples from nearly all of the principal soil formations of the State have been analyzed chemically and found to contain not less than from 5 to 10 tons of each of the principal plant foods per acre to a depth of one foot, which insures an abundance of food material for crops to feed on. The soils do not differ in their chemical composition sufficiently to account for the difference in their agricultural value.

The minerals composing the soil are broadly classed as gravel, sand, silt and clay, according to the size of the individual particles. These particles lie close together with small spaces between them, in which water is held in a film around the grains by a tension or capillary power. When the soils are nearly dry the film of water is very thin, and conversely when the soil is wet the film of water is much thicker, but the spaces are seldom completely filled with water in a well-drained soil, as there would then be no circulation of air within the soil, which is necessary for the full development of agricultural plants. The water moves either up or down in the soil as it is needed, according to capillary laws and the laws of gravity. The smaller the spaces the greater the force which moves the water; but the greater the friction and the slower the movement and conversely the larger the spaces, the less the power to hold or to move the water and the less the friction. Plants require a constant supply of water, but rain only falls on an average at intervals of a week or ten days, and the soil has to conserve and regulate this water supply so as to maintain a constant supply for the plants throughout the season.

Our different soil formations are very different in texture; some are very sandy, and others have but little sand in their composition, being composed mostly of clay and of fine silt. The sandy soils are coarse and open in texture and allow the rain to pass through them very readily. These soils have little power of maintaining this moisture for the plants, or of pulling it up from below to replace that lost by evaporation or used up by plants. The stiff clay soils, on the contrary, offer a great resistance to the rainfall, so that it moves down through them very slowly, and the soils have much more power of drawing it up again as needed than the sandy soils have. As a rule, these clay soils have four or five times as much moisture as the sandy soils, although the amount of rainfall, which is the source of supply of the water, is the same over both soils. If two plants were treated as differently as this in a greenhouse, one being given four or five times as much water as the other, the development of the plants would be very different; or, if during one season there were four or five times as much rainfall as in the preceding year, the effect on the crops would be greater than could be expected from any application of fertilizers.

It is an every-day matter for gardeners and florists to regulate the kind of development and the time of ripening of the crop or the flowering of plants, by judicious control of the conditions of moisture and heat. They can, indeed, force the plants to flower or fruit at will, or they can prolong the growth of the plants and prevent or greatly retard the maturity and induce a large leafy development. Any decrease in nutrition (water, air or mineral food) during the period of growth tends to check the growth of the vegetative parts of the plant and favors the production of fruit and the early maturity of the plant.

It can be shown that even with the same temperature and the same rainfall, where soils differ as much in texture as these do, the temperature and amount of moisture they can maintain for a crop will differ much more widely than these climatic conditions in widely different sections of the country, and these wide differences of climatic conditions *within* the soil have a marked effect upon the development, yield, quality and time of ripening of the crop. For instance, if a heavy limestone soil adjoined some of the light truck soil, and under precisely the same temperature and rainfall, we would find that the temperature in the light sandy soil at noon of a hot summer day was 10° or 15° *cooler* than in the heavy limestone soil, and it would be found that the clay land contained from 18 to 22 per cent of moisture, while the sandy land contained only 5 or 6 per cent of water.

The difference in the agricultural value of the soils of this State and their adaptation to the different crops are not due so much to the chemical composition of the soils and the amount of plant food they contain, as to their difference in texture and their relation to moisture and heat.

The following is a brief description of the most important soil formations in the State, with the exception of those
on the Eastern Shore, which have not yet been carefully
studied.   The soils are arranged about in the order of their
relative agricultural value.

1. **Pine Barrens,** *Pliocene or Lafayette Formation.*—The
prevailing soils of the Lafayette formation are a coarse
sand.   The subsoil contains only about 3 per cent of clay.
It has approximately 1,600,000,000 grains of sand and clay
in one gram.   These grains have about 496 square centimeters of surface for water to act on in dissolving food
material and for roots to feed on.   This is equivalent to
about 23,940 square feet of surface area in one cubic foot
of soil.

The grains are so large and there are relatively so few
of them in this soil that the lands are coarse and sandy,
and are so little retentive of moisture that they are not
able to maintain over 3 or 4 per cent of moisture, which
is too little for any agricultural crop.   At present these
lands are nearly valueless and are left out as pine barrens.
They would make the very earliest truck lands, however,
as crops would be forced to a very early maturity; and with
the intense system of cultivation which prevails in truck
farming, and when this part of the State is opened up and
good transportation facilities are offered, these will probably be the most valuable lands in the State for early truck.
Many of these lands also would bring a very fine grade of
tobacco for flue-curing.

2. **Early Truck and Fruit Lands,** *Pleistocene or Columbia,*
Eocene and Cretaceous Formations.—The early truck lands

of the Columbia formation, forming the river necks along the Bay shore and covering extensive areas on the Eastern Shore, have from 4 to 10 per cent of clay in the subsoil. The Eocene and Upper Cretaceous formations contain rather more than this, or from 8 to 15 per cent of clay, and are better adapted to fruit.

These light sandy lands along the Bay shore are altogether too light in texture for wheat, and with good treatment they would not bring over 5 or 10 bushels of wheat per acre. They are admirably adapted, however, to truck and fruit, and this industry has recently grown to very large proportions. These lands are so light in texture that they will only maintain, on an average, about 5 or 6 per cent of moisture for the crop, when the heavier soils in the western part of the State would have from 18 to 22 per cent. These drier conditions force the crop to an early maturity. Vegetables ripen on these early truck lands at least two or three weeks earlier than on any other soils of the State and bring a good market price.

These lands were formerly considered the very poorest lands in the State, and even now they have little value for general agricultural purposes when they are too remote from quick and easy transportation to enable truck to be grown on them. As a consequence, large areas of these lands are lying idle, awaiting improved transportation facilities. Many of these idle lands have about the same texture as the bright tobacco lands of North Carolina, and they will probably raise a fine grade of this tobacco for flue-curing.

The lighter soils are more valuable for early truck because they force the crops to an early maturity, so that they bring a good market price, while the heavier soils in this locality are better suited to fruit. Tomatoes, cabbage, and many other vegetables do better on the heavier lands and yield more per acre than on the lighter soils, but the crops are not so early and consequently do not bring so good a price. Tomatoes, for example, ripen at least a week earlier in a soil where there is only 5 per cent of clay in the subsoil, than they do on land having 8 per cent of clay, and this difference in time of marketing the crop materially increases the value, as the earlier the crop is matured the less competition there will be from other parts of the State and the better price it will bring in the market. Every effort is made to have the early truck mature at the earliest possible moment, and no reasonable expense is spared to attain this end. The whole value of these truck lands lies in the possibility of producing these vegetables earlier than they can be produced elsewhere in the State, so that the value of these truck lands does not depend so much upon the amount of crop they will produce per acre as upon the time the crop matures. Two or three days difference in the marketing of a crop in this trucking business may make the difference between a brilliant success and a failure.

3. Tobacco, Wheat and Grass Lands of Southern Maryland, *Miocene or Chesapeake Formation.*—There are three grades of land in the Chesapeake formation. The best tobacco lands of southern Maryland have from 12 to 18 per cent of clay in the soil, and on an average about 15 per cent. These lands are rather too light for profitable wheat production, but they make the finest grade of tobacco produced in Maryland.

Wheat and tobacco are commonly grown on the same land in rotation periods of two or three years, but the best wheat lands are too heavy in texture for the finer grades of tobacco, as the leaf is coarse and sappy and does not take on color in curing. These finer tobacco lands, however, make a very fine grade of tobacco; it has a good yellow color and is very mild and suitable for pipe-smoking. The tobacco is strictly an export tobacco, being sent principally to France and Holland.

The wheat lands of southern Maryland contain from 18 to 25 per cent of clay in the subsoil, and on an average about 20 per cent. They contain approximately 9,000,000,-000 grains of sand and clay in one gram, and these grains have about 2000 square centimeters of surface.

These lands are sufficiently retentive of moisture to give fair yields of wheat, but they are near the limit of profitable wheat production, for soils lighter in texture than these are too light for wheat with the prevailing amount and distribution of rainfall. Subsoils having less than 18 per cent of clay, or approximately 9,000,000,000 grains of sand and clay in one gram, are, as a rule, rather too light in texture for profitable wheat production in this locality, as they are not sufficiently retentive of moisture and do not maintain a sufficient water supply for the plants. With the present prices of wheat it would be too costly to attempt to fill up the spaces in a lighter soil with organic matter, or to rearrange the grains of sand and clay so as to make the soil more retentive of moisture. The conditions in the lighter soils are not constant, and in unfavorable seasons the plants suffer and the yield is very small.

These wheat lands are too light in texture for grass or for permanent pasture, and, on the other hand, they are too heavy in texture and maintain too much moisture for the better grade of tobacco grown in southern Maryland, for the leaf produced is coarse and sappy, cures green and does not take on color, and brings a very low price.

A soil must have at least 25 or 30 per cent of clay in the subsoil to make a good grass land, unless there is more organic matter present than our Maryland soils usually contain, or unless the grains of sand and clay are differently arranged from what is generally the case here. There is a considerable area of land in this Chesapeake formation which has 30 per cent of clay and over, which is well suited to grass. These grass lands also make the very finest wheat lands in that portion of the State.

4. **Wheat and Corn Land,** *Pleistocene or Columbia Formation.*—The fertile terraces bordering the Potomac and Patuxent rivers and their tributaries, and the Columbia formation where it occurs at high levels in other parts of the State (the fluvial and interfluvial phases), have from 20 to 30 per cent of clay in the subsoil, and on an average about 25 per cent. These lands are sufficiently retentive of moisture to make excellent wheat and corn lands. Some of these lands are known to have been cultivated for upwards of 200 years, yet they show no signs of deterioration.

5. **Grass, Wheat and Corn Lands,** *Gabbro, Gneiss and Phillite Formations.*—These three formations are so nearly alike in texture and in agricultural value that they may be described together. They contain on an average about 30 per cent of clay in the subsoil, and have about 14,400,-000,000 grains of sand and clay in one gram. The gabbro lands are, as a rule, rather closer in texture than the

others, and are rather better for general agricultural purposes for this reason. The land is sufficiently retentive of moisture to make very fine wheat and grass lands, and in favorable seasons and with good treatment from 20 to 30 bushels of wheat can be produced on this land. The land is well adapted to grazing purposes, and large numbers of store cattle are annually fattened for market on the gabbro soils of Harford County.

There is a much larger area of gneiss in the State than of gabbro. As a rule, the soil is sufficiently retentive of moisture to maintain good pasturage, and it makes excellent wheat and corn lands. A large number of store cattle are fattened on the heavier soils, and the dairy interests are very extensive. Some of the soils are rather too light for profitable wheat production, and these are admirably adapted to truck and vegetables, but the crops ripen so late that they come into competition with crops from other parts of the State, and they do not bring as good prices as crops from the lighter truck lands of southern Maryland and the Eastern Shore. Large crops of tomatoes and corn are raised on the lighter soils of the gneiss formation, for canning, and this interest has replaced the cultivation of wheat to a large extent.

The phillite formation covers the northern part of Harford, Carroll, Howard, and Montgomery, and the eastern part of Frederick counties. It has, as a rule, about the same texture and the same agricultural value for grass, wheat and corn as the two formations just described. Tobacco is grown to a limited extent on newly cleared phillite lands, but it is a very much larger and heavier leaf and has altogether a different texture from that grown on the lighter soils of southern Maryland.

6. **Grass and Wheat Lands,** *Triassic and Catskill Red Sandstone Formations.*—These subsoils contain about 35 per cent of clay, and have about equal agricultural value. The triassic red sandstone is locally known as the "red lands" of Carroll and Frederick counties. It is sufficiently retentive of moisture to make admirable grass and wheat lands. It adjoins the Trenton limestone, which is the strongest type of grass land in the State. In favorable seasons and with good treatment these triassic red sandstone soils will make about as much wheat per acre as the adjoining limestone lands, but the crop is never as safe nor as certain, for the soil is not as heavy in texture as the limestone land; it is not as retentive of moisture, and the crop is much more affected by unfavorable seasons and by extremes of wet and dry weather. These lands, like the limestone lands, are greatly benefited by an application of lime. They are easier to work than the limestone lands, but, on the other hand, they cannot stand such hard farming as the heavier limestone soils can. The Catskill red sandstone forms some very fertile valleys in Garrett and Allegany counties. They are very strong clay lands, very retentive of moisture, and are admirably adapted to wheat and grass.

7. **Grass and Wheat Land,** *Trenton Limestone.*—This is the strongest and finest type of grass and wheat land in the State. The subsoil contains from 40 to 50 per cent of clay, which makes it very retentive of moisture. There is, on an average, about 45 per cent of clay, and approximately 22,000,000,000 grains of sand and clay in one gram of this subsoil, which divides up the empty space very much, and the rainfall has to pass down through the innumerable little

passages between these grains. The grains of sand and clay in one cubic foot of this subsoil have no less than 158,000 square feet of surface for water to act on in dissolving food material and for roots to feed on. In a cubic foot of this subsoil there is therefore no less than 3½ acres of surface exposed to the action of the water and roots. This enormous extent of surface makes it possible, of course, for the plants to extract a considerable amount of food material from the soil. The large number of grains in this subsoil makes a very fine and close-textured, stiff clay, which is very retentive of moisture, although it is also well drained. Good crops of grass and wheat are assured in all ordinary seasons, and with good treatment from 30 to 40 bushels of wheat per acre can be produced on this land.

These soils are the impurities originally contained in the limestone rock, which have been left behind as the lime has been dissolved and carried off by water. There is, of course, a very small amount of impurities in the limestone rock, and after the large amount of lime has been dissolved the impurities settle, and, as a consequence, the limestone soils are nearly always valley lands with ridges on either side formed of rocks which were much less soluble than the limestone. Another important fact is that the lime is in the form of a carbonate which is readily soluble in water containing carbonic acid gas in solution, whereas the lime in most ordinary soils is in the form of sulphate or silicate, either of which is much less soluble in water than the carbonate, so it happens that, strange as it may seem, these limestone soils are frequently deficient in lime, and there is no class of soils in the State which is more benefited by an application of lime than these same soils resulting from the disintegration of the limestone rocks. It is very frequently the practice in these limestone regions to get out the rock and burn it in kilns and spread it directly on the land from which it came.

Several of the other soil formations of the State deserve notice, although too little work has been done in them to warrant any detailed description.

The Cambrian sandstone is where the mountain peaches have been so successfully grown. The soil contains considerable clay, but it is filled with fragments of thin pieces of sandstone.

The very fertile soils of the famous Middletown Valley have not been studied, as the geology of that region has only been very recently worked out.

The soils of the mountainous and coal areas in Washington, Allegany, and Garrett counties have not been studied in much detail. There are large areas of these lands which have no agricultural value, but there are also fertile valley lands.

The Potomac formation crossing the State from Washington through Baltimore to the Delaware line is one of the poorest sections of the State. The prevailing soils are vari-colored clays containing from 40 to 50 per cent of clay, and these should make very fertile lands, but on account of the arrangement of the grains of sand and clay the soils are very close and so impervious to water that they are not suited to agricultural crops. The valley lands, however, where these clays are overlaid by the Columbia, give fertile wheat and corn lands.

The soils of the Eastern Shore have not yet been studied in sufficient detail to be described here.

THE CLIMATE.

Although the climate in general is what is known as continental, it is greatly modified in the eastern portion of the State by the ocean and the Chesapeake Bay, and in the extreme southeast becomes almost oceanic insular, or surrounded as the land is on nearly all sides by water.

The State of Maryland will naturally fall into the four following climatic divisions:

EASTERN MARYLAND, } *Coastal Plain.*
SOUTHERN MARYLAND }
NORTHERN-CENTRAL MARYLAND—*Piedmont Plateau.*
WESTERN MARYLAND—*Appalachian Region.*

The following tables show the monthly, seasonal, and annual means of temperature and precipitation for the State and its different regions, and the accompanying charts graphically exhibit the distribution over the State of the annual and seasonal means:

Monthly and Annual Means of Temperature.

	January.	February	March.	April.	May.	June.	July.	August.	September.	October.	November.	December.	Year.
STATE	32.8	34.8	39.6	51.7	62.6	72.5	75.8	74.3	66.9	54.7	44.0	35.5	53.8
Eastern Maryland	34.8	36.1	40.5	52.6	62.1	72.8	75.8	74.8	67.5	56.5	45.3	37.3	54.5
Southern Maryland	35.3	37.4	42.3	53.4	63.9	74.1	77.7	75.7	68.6	56.6	46.5	37.6	55.6
North-Central Maryland	30.7	33.9	38.2	50.9	63.5	72.8	75.7	72.4	65.8	54.2	42.8	34.0	53.0
Western Maryland	30.5	31.6	37.2	49.8	60.7	70.3	73.8	74.2	65.8	51.4	41.2	33.0	52.0

Monthly and Annual Means of Precipitation.

	January.	February.	March.	April.	May.	June.	July.	August.	September.	October.	November.	December.	Year.
STATE..............	3.31	3.07	3.92	3.75	4.21	3.72	4.11	3.77	3.67	2.75	3.22	2.69	42.43
Eastern Maryland......	3.51	3.22	4.06	4.04	4.29	3.18	4.78	3.78	3.39	3.04	2.70	2.67	42.66
Southern Maryland.....	3.20	3.51	4.20	4.11	4.40	3.70	4.42	3.84	3.50	2.86	4.11	2.60	44.75
North-Central Maryland	3.50	3.10	4.39	3.62	4.06	3.48	4.45	3.98	4.03	2.74	3.25	3.13	43.73
Western Maryland......	3.01	2.45	3.02	3.23	4.08	4.53	2.77	3.48	3.46	2.35	2.82	2.35	38.55

Seasonal Means of Temperature and Precipitation.

	MEAN TEMPERATURE.				MEAN PRECIPITATION.			
	Spring.	Summer.	Autumn.	Winter.	Spring.	Summer.	Autumn.	Winter.
STATE......................	51.2	74.0	55.0	34.4	12.88	11.60	9.64	9.31
Eastern Maryland.............	51.7	74.5	55.8	36.1	12.39	11.74	9.13	9.40
Southern Maryland............	53.1	75.5	57.2	36.9	12.71	11.96	10.77	9.31
North-Central Maryland........	50.6	73.5	54.3	33.1	12.07	11.91	10.02	9.73
Western Maryland.............	49.4	72.7	52.7	31.7	10.33	10.78	8.63	8.81

Temperature.

The coldest month is January, with an average mean temperature for the State of 32.8°, while the warmest month is July, with an average mean temperature of 75.8°, a difference of 43°. The greatest changes in mean temperature take place in the Spring and Autumn months, while those in Summer and Winter are very small.

The mean annual temperature of the western division is 52°, while that of the southern is 55.6°, a difference of 3.6°. The mean annual temperature has a much greater range, however, when the extremities of the State are compared. The isothermal line of 50° passes through Garrett and Allegany counties and bends down along the high ridge of the Piedmont Plateau into Carroll and

Baltimore counties, while the isothermal line of 58° crosses Worcester and Somerset counties to the Virginia shore of the Chesapeake. There is thus a difference of over 8° in the annual means between the extreme northern and western and the extreme southern portions of the State.

The seasonal isothermal lines indicate a still wider range in mean temperature between the western and southeastern portions of the State. In spring it ranges from 56° to 44°, a difference of 12°, in summer from 77° to 69°, a difference of 8°, in autumn from 60° to 50°, a difference of 10°, and in winter from 40° to 27°, a difference of 13°.

Eastern Maryland. This portion of the State, designated as the eastern division of the Coastal Plain, is deeply indented by tidal estuaries and bordered by the ocean, its temperature being much modified by the surrounding water.

The southern portion of the area has a mean annual temperature of 58°, the highest in the State. In passing to the northward the temperature changes at first rapidly, the isothermal line of 57° and 56° following at short intervals. The greater portion of the eastern division, however, is found between the isothermal lines of 56° and 54°, while in the extreme north the temperature again changes rapidly, the isothermal lines of 53° and 52° following each other at short intervals. The extreme range in the mean annual temperature is thus found to be 6°.

The mean seasonal variations between the southern and northern portions of the region are also distinctly marked. As in the case of the annual means, the isothermal lines do not succeed each other generally at regular intervals.

The mean temperature for spring ranges from 50° in the north to 56° in the south. The greater portion of the region is found, however, between the means of 51° and 53°.

In summer there is very little range in mean temperature between the northern and southern portions of the district. The entire region lies between the isothermals of 74° and 76°.

In autumn the range in mean temperature is the same as in spring, amounting to 6°. Although the extremes are found between 54° and 60°. the greater portion of the region lies between 55° and 57°.

The greatest difference in mean temperature is found in winter. The variation is then 9°, and the mean temperature ranges from 31° in the north to 40° in the south. There is a much more gradual change than at other seasons, the isothermals being found approximately equidistant from one another.

Southern Maryland. The southern portion of the State has been described as the western division of the Coastal Plain. The surface of the land is somewhat higher and more broken than in Eastern Maryland, but is still low and flat. On account of this general uniformity throughout the area, together with its limited extent from north to south, the variations in mean temperature are not very striking. The annual mean seldom exceeds that of Baltimore, which is 55.6°, by more than 2°, while Leonardtown and several other places have almost the same average temperature. At a few points, owing to local causes, the mean annual temperature is even lower.

With the exception of the winter temperature, the mean seasonal temperatures show very slight variations,

Graphical Representation of Mean Temperatures in the Four Climatic Divisions of Maryland.

WESTERN MARYLAND
NORTHERN-CENTRAL MARYLAND
EASTERN MARYLAND
SOUTHERN MARYLAND

seldom reaching more than two degrees. In spring the region is crossed by the isothermal line of 53° and 54°, in summer of 75°, 76° and 77°, in autumn of 57°, 58°, 59° and 60°, the two latter, however, only touching the southern portion of St. Mary's county. In winter, on the other hand, variations of four or five degrees are found, the isothermal lines of 36°, 37°, 38°, 39° and 40° succeeding one another at very nearly equal intervals.

The interior portion of the country is warmer during the spring, summer and autumn months, but cooler during the winter.

Northern-Central Maryland. The hilly country bordering the Coastal Plain upon the east and lying mainly in the Piedmont Plateau is here referred to under the name of Northern-Central Maryland.

The mean annual temperature of the region ranges from 50° to 55°. The coldest portions are found along the higher land of the Piedmont belt which culminates in Parr's Ridge. The Frederick valley is considerably warmer, corresponding in this respect with the eastern slope in Montgomery and Howard counties.

The mean seasonal temperatures have the same general relations to the topography as the annual temperatures. The high-central portion of the Piedmont area is at all seasons several degrees colder than the eastern slope or the Frederick valley. The spring means vary from 48° to 53°, the summer from 69° to 75°, the autumn from 52° to 57°, and the winter from 29° to 36°, which indicates a slightly greater range in temperature in winter and summer than in the spring and autumn.

Western Maryland. The portion of the State which is here considered under the name of Western Maryland has been described as the Appalachian Region. It consists of parallel ranges of mountains, with deep valleys, which drain chiefly into the Potomac river. The mountains reach 3000 feet and more in altitude, and in the west rise from a high plateau, which declines gradually beyond the limits of the State.

As might be anticipated, there is a general lowering of the temperature throughout the entire district.

So far as conclusions can be drawn from the records of temperature, which are not altogether satisfactory, the valleys are warmer than the mountains. This is best seen in the Hagerstown valley, where the isothermals invariably bend to the westward. In the smaller valleys few continuous observations have been taken, while, practically, none are recorded from the mountains, with which comparisons may be made.

There is a slight decrease in the mean annual temperature in passing from the eastern to the western portions of the region. The range is from 50° to 53°, making a difference of 3°.

This is shown more distinctly in the case of the seasonal means, particularly in the spring and winter. In the spring the mean temperature varies from 44° to 52°; in summer, from 70° to 75°; in autumn, from 50° to 54°; in winter, from 27° to 34°, which shows a greater variation in spring and winter than in summer and autumn.

Precipitation.

The atmospheric precipitation in Maryland occurs both as rain and snow. There is no portion of the State in which either is entirely wanting, although the snowfall is far less in the eastern and southern districts than in the northern and western. The snowfall never fails completely even in the warmest winters, although it may be reduced to insignificant proportions.

The precipitation is more or less equally distributed throughout the months, when the means for a long term of years are taken into consideration, although wet and dry periods characterize the seasons of a single year, causing variations from the normal. A certain constant increase in the mean precipitation is found to occur in the spring and late summer, with a corresponding decrease in the autumn and winter.

The western portion of the State has a less amount of annual precipitation than the eastern. A heavy rainfall characterizes the region which lies to the east of Catoctin Mountain, the easterly winds, as they reach the highlands, precipitating their moisture in the Frederick valley and over the western slope of the Piedmont Plateau. The eastern slope of the Piedmont Plateau has again less precipitation.

The western portion of the Coastal Plain has a much drier climate than the eastern, although numerous local exceptions appear. For example, the western shores of the Chesapeake have relatively much greater precipitation than the eastern, which makes the average precipitation for Southern Maryland exceed that for Eastern Maryland. The central and western portion of Eastern Maryland has a much greater rainfall than the area bordering the Atlantic.

The precipitation generally accompanies the areas of low pressure which traverse the country from west to east, and pass to the north of Maryland. It commonly occurs on their eastern front during the prevalence of easterly winds.

Winds.

The prevailing winds in Maryland are northwesterly. In the eastern and southern portions of the State they are frequently southerly during the summer months, but during the remainder of the year are more from the west and northwest. In the mountainous regions of Western Maryland the winds are more constantly from the northwest and west throughout the year. In general the westerly direction of the wind veers more and more to the southerly in passing from the inland mountainous region toward the coast.

The barrier of highland which stretches across the western portion of the State protects the entire area of Maryland from the destructive effects of tornadoes and violent storms so frequent farther west.

Along the shore line of the State during the warmer months there are inflowing currents of air, or sea-breezes, which moderate the temperature of the land for some distance from the coast. They generally blow from midday till sundown, and are due to the heated atmosphere over the land rising and thus causing the cooler air over the water to flow in to take its place.

Humidity.

The capacity of the atmosphere to hold moisture varies, but vapor of water is always present in greater or less amounts. When the atmosphere is near saturation the air is moist, but when it is capable of taking more water it becomes dry in proportion to the amount which can thus be taken. If the saturated state is taken as the standard of comparison, or 100, then the relative amount of moisture can be indicated by percentage.

Observations have been recorded at comparatively few points, so that reliable means are difficult to obtain. The following table gives the relative humidity of a few stations during the year 1892:

STATIONS.	January.	February.	March.	April.	May.	June.	July.	August.	September.	October.	November.	December.	Year.
Baltimore	79	78	73	65	69	77	72	73	72	67	76	74	73
Barron Creek Springs ..	85	87	84	72	72	80	78	82	78	71	77	84	79
McDonogh	80	80	74	69	66	75	73	82	80	72	76	70	75
Washington, D. C	74	73	70	65	70	75	76	73	74	68	70	74	72

From 1871 to 1892 the mean relative humidity in Baltimore has been as follows:

1871 to 1892.	January.	February.	March.	April.	May.	June.	July.	August.	September.	October.	November.	December.	Year.
Baltimore	70	65	64	61	65	68	68	70	74	68	70	68	67.6

Barometric Pressure.

The variations in barometric pressure are not very great in the more populous portions of the State. Since none of the larger towns are situated at a height of even 1,000 feet above sea level, the variations in the mercury column due to elevation would not, at ordinary temperatures, exceed one inch. Even the highest ranges of the western portion of the State would show a difference of but little over three inches. In recording barometric observations, however, it is customary to make corrections by reducing the readings to a common datum, which is that of sea level. The most important variations in the barometric pressure are due to the passage of areas of low pressure, few of which, however, take their track directly across the State. Most of them pass to the north of the confines of Maryland. Their coming is generally accompanied by rainfall, and is preceded by a rise and followed by a fall in temperature. Barometric observations have been taken continuously at only a few points in the State, and no important general conclusions can be drawn from the records. The mean monthly barometric readings for Baltimore from 1871 to 1892 are given in the following table:

January.	February.	March.	April.	May.	June.	July.	August.	September.	October.	November.	December.	Year.
30.06	30.14	29.99	29.98	29.98	30.00	29.90	30.00	30.01	30.06	30.10	30.14	30.03

MONTHLY SUMMARY OF THE WEATHER
FOR 1892 AND 1893.

JANUARY, 1892.

Temperature (degrees).—Mean monthly, 31.8; highest monthly mean, 35, at Cumberland (1) and Barron Creek Springs; lowest monthly mean, 28, at Darlington and Jewell; highest temperature, 70, at Barron Creek Springs and Kirkwood, Del., on the 14th; lowest temperature, 0, at Boettcherville, on the 8th; greatest local monthly range, 63, at Charlotte Hall; least local monthly range, 42, at New Market; mean monthly range, 54.0. Mean maximum temperature, 39.5; mean minimum temperature, 24.4.

Precipitation (in inches).—For entire territory covered, 4.45; greatest amount, 6.54, at Fallston; least amount, 1.87, at Taneytown.

FEBRUARY, 1892.

Temperature (degrees).—Monthly mean, 35.4; highest monthly mean, 38, at Cumberland (1) and Easton; lowest monthly mean, 28, at Darlington; highest temperature, 64, at Cumberland (1); lowest temperature, — 8, at Boettcherville, on the 6th; greatest local monthly range, 68, at Boettcherville; least local monthly range, 38, at Leonardtown; mean monthly range, 49.0. Mean maximum temperature, 43.2; mean minimum temperature, 29.1.

Precipitation (in inches).—Average for entire territory covered, 2.39; greatest amount, 4.07, at Solomon's; least amount, 1.56, at Great Falls.

MARCH, 1892.

Temperature (degrees).—Monthly mean, 37.3; highest monthly mean, 40, at Cumberland (1); lowest monthly mean, 34, at Fallston and Mt. St. Mary's; highest temperature, 67, at Cumberland (1) on the 25th; lowest temperature, 0, at Charlotte Hall, on the 5th; greatest local monthly range, 58, at Charlotte Hall; least local monthly range, 31, at Denton; monthly mean range, 44.3. Mean maximum temperature, 45.4; mean minimum temperature, 29.2.

Precipitation (in inches).—Average for entire territory covered, 4.87; greatest amount, 7.20, at Baltimore; least amount, 2.50, at Boettcherville.

APRIL, 1892.

Temperature (degrees).—Mean monthly, 50.0 ; highest monthly mean, 54.0, at Easton ; lowest monthly mean, 48.0, at Kirkwood, Del. ; highest temperature, 87, at Cumberland (2), on the 4th ; lowest temperature, 23, at Boettcherville, on the 12th ; greatest local monthly range, 65, at Woodstock ; least local monthly range, 44, at McDonogh and the Receiving Reservoir, D. C. Mean monthly range, 51.8 ; mean maximum, 59.8 ; mean minimum, 42.1.

Precipitation (in inches).—Average, 3.89 ; greatest amount, 6.68, at Barron Creek Springs ; least amount, 2.03, at McDonogh.

Wind.—Prevailing direction, northwest. Total movement in miles, Baltimore, 6433 : Norfolk, Va., 7421 ; Washington, D. C., 5705.

MAY, 1892.

Temperature (degrees).—Mean monthly, 63.1 ; highest monthly mean, 64.8, at Easton ; lowest monthly mean, 60.2, at Fallston ; highest temperature, 90, at Seaford, Del., on the 4th ; lowest temperature, 35, at Boettcherville, on the 9th ; greatest local monthly range, 51, at Boettcherville ; least local monthly range, 28, at Jewell. Mean monthly range, 41.1 ; mean maximum, 72.4 ; mean minimum, 53.6.

Precipitation (in inches).—Average, 4.37 ; greatest amount, 6.35, at Baltimore ; least amount, 2.16, at Frederick.

Wind.—Prevailing direction, southwest. Total amount in miles, Baltimore, 5987 ; Norfolk, Va., 7363 ; Washington, D. C., 5252.

JUNE, 1892.

Temperature (degrees).—Mean monthly, 75.3 ; highest monthly mean, 78.7, at Cumberland (1) ; lowest monthly mean, 71.2, at Seaford, Del. ; highest temperature, 98, at New Market, on the 24th ; lowest temperature, 49, at Seaford, Del., on the 11th ; greatest local monthly range, 46, at Seaford, Del. ; least local monthly range, 19, at Jewell. Mean monthly range, 36.9 ; mean maximum, 83.8 ; mean minimum, 66.9.

Precipitation (in inches).—Average, 3.78 ; greatest amount, 10.08, at Cumberland (2) ; least amount, 1.42, at the Distributing Reservoir, D. C.

Wind.—Prevailing direction, southwest. Total movement in miles, Baltimore, 5635 ; Norfolk, Va., 6776 ; Washington, D. C., 4550.

JULY, 1892.

Temperature (degrees).—Mean monthly, 75.8; highest monthly mean, 78.9, at Kirkwood, Del.: lowest monthly mean, 72.9, at Cumberland (2); highest temperature, 102. at Kirkwood, Del., on the 27th and 28th ; lowest temperature, 50, at Boettcherville, on the 19th : greatest local monthly range, 50, at Boettcherville ; least local monthly range, 27, at Solomon's. Mean monthly range, 41.1 : mean maximum, 85.3 ; mean minimum, 66.2.

Precipitation (in inches).—Average, 3.81 : greatest amount, 6.40, at the Distributing Reservoir. D. C.: least amount, 1.10, at Boettcherville.

Wind.—Prevailing direction, southwest. Total movement in miles, Baltimore, 4451 : Norfolk, Va., 5457 : Washington, D. C., 3521.

AUGUST. 1892.

Temperature (degrees).—Mean monthly. 75.9 ; highest monthly mean, 79.7, at Kirkwood, Del. ; lowest monthly mean, 72.1, at Fallston ; highest temperature, 101, at Cumberland (1), on the 9th ; lowest temperature, 56, at Boettcherville, on the 20th, and at Edgemont, on the 14th ; greatest local monthly range, 43, at Edgemont ; least local monthly range, 15, at Jewell. Mean monthly range, 34.4 : mean maximum, 85.5 ; mean minimum, 66.5.

Precipitation (in inches).—Average, 1.96 ; greatest amount, 4.10, at Fallston : least amount, .58, at Taneytown.

Wind.—Prevailing direction. northwest. Total movement in miles, Baltimore, 4536 : Norfolk, Va.. 4470 ; Washington, D. C., 3613.

SEPTEMBER, 1892.

Temperature (degrees).—Mean monthly. 65.9 ; highest monthly mean, 69.5, at Kirkwood, Del. ; lowest monthly mean, 63.2, at New Market ; highest temperature, 98, at Edgemont. on the 4th and 5th ; lowest temperature, 40. at Boettcherville, on the 8th ; greatest local monthly range, 52, at Edgemont : least local monthly range, 25, at Jewell. Mean monthly range, 37.9 ; mean maximum, 65.9 ; mean minimum, 56.3.

Precipitation (in inches).—Average, 3.05 ; greatest amount, 5.82, at Frederick ; least amount, 1.75. at Solomon's.

Wind.—Prevailing direction, southeast. Total movement in miles, Baltimore, 4867 : Norfolk, Va., 5311 : Washington. D. C., 3908.

OCTOBER, 1892.

Temperature (degrees).—Mean monthly, 54.9; highest monthly mean, 61.9, at Cambridge; lowest monthly mean, 48.4, at Oakland; highest temperature, 88, at Cambridge, on the 3rd; lowest temperature, 13, at Sunny Side, on the 31st; greatest local monthly range, 68, at Sunny Side; least local monthly range, 47, at Solomon's; mean monthly range, 53.8. Mean maximum temperature, 65.3; mean minimum temperature, 45.6.

Precipitation (in inches).—Average, 4.42; greatest amount, 6.40, at Valley Lee; least amount, 2.59, at Darlington.

NOVEMBER, 1892.

Temperature (degrees).—Mean monthly, 42.9; highest monthly mean, 46.8, at Leonardtown; lowest monthly mean, 40.0, at Boettcherville; highest temperature, 77.0, at Edgemont, on the 4th; lowest temperature, 15.0, at Edgemont, on the 24th; greatest local monthly range, 62, at Edgemont; least local monthly range, 34, at Denton. Mean monthly range, 48.3; mean maximum, 51.2; mean minimum, 35.9.

Precipitation (in inches).—Average, 4.31; greatest amount, 6.60, at Seaford, Del.; least amount, 3.10, at Easton.

Wind.—Prevailing direction, northwest. Total movement in miles, Baltimore, 6567; Norfolk, Va., 7481; Washington, D. C., 5743.

DECEMBER, 1892.

Temperature (degrees).—Mean monthly, 32.3; highest monthly mean, 36.0, at Solomon's; lowest monthly mean, 29.4, at Kirkwood, Del.; highest temperature, 68.0, at McDonogh; lowest temperature, 6.0, at Boettcherville and Woodstock, on the 28th and 29th, respectively; greatest local monthly range, 58, at Boettcherville, Great Falls and Woodstock; least local monthly range, 42, at Kirkwood, Del. Mean monthly range, 51.8; mean maximum, 39.3; mean minimum, 26.3.

Precipitation (in inches).—Average, 2.24; greatest amount, 3.02, at the Distributing Reservoir, D. C.; least amount, 1.37, at McDonogh.

Wind.—Prevailing direction, northwest. Total movement in miles, Baltimore, 5567; Norfolk, Va., 5942; Washington, D. C., 4644.

JANUARY, 1893.

Temperature (degrees).—The month of January, 1893, will long be remembered for its accompaniment of extremely cold weather, which is well known not to have been at all local in character, nearly every section of the country having been invaded by a temperature very low in comparison with previous records. Probably not since January, 1856, has there been experienced in Maryland, the District of Columbia, and Delaware such a protracted period of severe weather. It is certain that not during the life of the Weather Bureau, which came into existence in 1870, has anything approaching a parallel been experienced. There have been lower temperatures in previous years, but they endured for a day or two only, being quickly succeeded by comparatively warm weather; the mean of several days, or of an entire month, places the recent frigid period far in the lead.

The mean temperature of the month at Baltimore was 25, and nothing approaches nearer to it than 29 in 1886 and in 1888. This is a decided difference in mean temperatures, especially between lowest mean temperatures. The mean of the month at Washington was still lower, 24, and the next lowest, 28, which occurred in 1881. A minimum temperature of 6 below zero was recorded at Washington on the 18th, while the lowest at Baltimore was 1 above zero, on the 16th.

Mean monthly, 22.9; highest monthly mean, 27.0, at Cambridge; lowest monthly mean, 13.1, at Sunny Side. Highest temperature, 58, at Cumberland (1), at Edgemont and Leonardtown, on the 27th; lowest temperature, 17 below zero, at Denton, and at Millsboro, Del., on the 17th and 18th, respectively. Greatest local monthly range, 73, at Denton; least local monthly range, 25, at Cambridge. Mean monthly range, 56.9. Mean maximum, 31.9. Mean minimum, 16.4.

Precipitation (in inches).—Average, 1.94; greatest amount, 3.50, at Sunny Side; least amount, 0.72, at Cumberland (2). The greatest fall of snow during the month, in Maryland, 36 inches, is reported by the observer at Oakland, Garrett Co. The next greatest fall, 13.7, was at Cambridge, Dorchester Co. Only 6 is reported from Cumberland, which seems remarkable considering the situation of that place. Boettcherville, close by, reports 12.5. Baltimore reports 8.1. The greatest depth of snow in Delaware was 17.0, at Seaford. The observer at Penn's Grove, N. J., reports 16.0, which probably approximates closely with the fall in Northern Delaware.

Wind.—Prevailing direction, northwest. Total movement in miles, Baltimore, 6510; Norfolk, Va., 6247; Washington, D. C., 5549.

FEBRUARY, 1893.

Temperature (degrees).—Mean monthly (for entire territory covered), 33.3 ; highest mean monthly, 38.9, at Cambridge ; lowest mean monthly, 24.4, at Sunny Side. Highest temperature, 70, at Denton ; lowest temperature, 5 below zero, at Boettcherville, on the 21st. Greatest local monthly range, 63, at Boettcherville ; least local monthly range, 43, at Salisbury. Mean monthly range, 52.4. Mean maximum temperature, 42.3; mean minimum temperature, 26.4.

Precipitation (in inches).—Average, 4.23 ; greatest amount, 5.45, at Woodstock ; least amount, 2.22, at Jewell. The greatest fall of snow during the month, in Maryland, 22 inches, is reported by the observer at Sunny Side, Garrett Co. The next greatest fall, 21 inches, was at Boettcherville, Allegany Co. Cumberland, close by, reports 20 inches. The least fall, 1.5 inches, is reported by the observer at Leonardtown, St. Mary's Co. Baltimore reports 11.7.

Wind.—Prevailing direction, northwest. Total movement in miles, Baltimore, 6628 ; Norfolk, Va., 7470 ; Washington, D. C., 5877. A severe wind storm occurred in Maryland on the 19th, the full particulars of which are given in the "Notes by Observers."

MARCH, 1893.

Temperature (degrees).—Mean monthly (for entire number of stations), 39.7 ; highest monthly mean, 44.4, at Cambridge ; lowest, 34.0, at Sunny Side ; highest temperature, 73.0, at Barron Creek Springs ; lowest, 6, at Sunny Side. Mean maximum temperature, 49.4 ; mean minimum, 31.8 ; greatest local monthly range, 64.0, at Sunny Side ; least, 42.0, at Great Falls. Mean monthly range, 50.4. The mean monthly temperature varied but slightly from the normal, being, probably, a little below it.

Precipitation (in inches).—Average, 3.32 ; the greatest amount, 4.14, fell at Cambridge, and the least, 1.00, at Cumberland. At Baltimore the total, 1.38, is smaller than in any previous year since the establishment of the station in 1871. Consulting the oldest records, only one instance can be found of a smaller precipitation total for the month at Baltimore, and that was 1.30 in 1822, as reported by Lewis Brantz. These old records extend from 1817 to 1837 and from 1836 to 1859. In a record of 50 years' observations at Ft. McHenry (from 1836), smaller amounts are found to have fallen in 1885 (1.24) and in 1858 (1.31).

The precipitation throughout the past month was well distributed, except during the last week, when none fell anywhere.

About .2 of the entire amount of precipitation was in the form of snow. 12.5 inches, the maximum amount, fell at Millsboro, Del., 12.3 at Cambridge, and 12 at Denton. At several stations in northern, central and western Maryland, none fell.

Wind.—Prevailing direction, northwest. Total movement in miles, Baltimore, 6904 ; Norfolk. Va., 7529 ; Washington, D. C., 5898.

APRIL, 1893.

Temperature (degrees).—Monthly mean (for entire territory covered), 52.4 ; highest monthly mean, 58.6, at Benedict ; lowest monthly mean, 46.2, at Sunny Side ; highest temperature, 92, at Boettcherville, on the 8th ; lowest temperature, 22, at Sunny Side, on the 24th ; greatest local monthly range, 59, at Sunny Side ; least local monthly range, 37, at Distributing Reservoir, D. C.. and Receiving Reservoir, D. C. Mean monthly range, 45.2 ; mean maximum temperature, 61.9 ; mean minimum temperature, 44.3.

Precipitation (in inches).—Average, 3.94 ; greatest amount, 6.75, at Oakland ; least amount, 2.56, at Jewell. The greatest fall of snow during the month, in Maryland, 3.5 inches, was reported by the observer at Sunny Side.

Wind.—Prevailing direction, southeast. Total movement in miles, Baltimore, 6604 ; Norfolk, Va., 8278 ; Washington, D. C., 5869.

MAY, 1893.

Temperature (degrees).—Monthly mean (for entire territory covered), 61.2 ; highest monthly mean, 66.2. at Cambridge ; lowest monthly mean, 53.2, at Sunny Side ; highest temperature, 92, at Boettcherville. and Millsboro, Del., on the 23rd ; lowest temperature, 33. at Sunny Side, on the 10th ; greatest local monthly range, 55, at Millsboro, Del. ; least local monthly range, 37, at Cambridge ; mean monthly range, 46.2. Mean maximum temperature, 71.8 ; mean minimum temperature, 51.7.

The month was much colder than usual. At Baltimore, with an average of 61°, it was the coldest May since the establishment of the station, in 1871. with the exception of the month in 1882, the mean temperature of which was 59°.

It may be conjectured that the low temperature was due, in part, to the slowness with which a cold winter relaxed its grip, and in part to the number and preponderance of the storms from the southwest, which gave the cold, heavy air, over the still frozen northern regions, an opportunity of flowing southward in unusual volume.

Precipitation (in inches).—Average, 4.78 ; greatest amount, 6.60, at Fenby ; least amount, 3.06, at Milford, Del. Snow, to the amount of a trace only, was reported from the following stations : Baltimore, Glyndon, Oakland, and Sunny Side.

There was about the average amount of rainfall, and it was evenly distributed over the territory and throughout the month.

Wind.—Prevailing direction, southeast. Total movement in miles, Baltimore, 6588 ; Norfolk, Va., 6949 ; Washington, D. C., 5167.

JUNE, 1893.

Temperature (degrees).—Monthly mean (for State), 72.3 ; highest monthly mean, 76.6, at Cambridge ; lowest monthly mean, 65.0, at Sunny Side ; highest temperature, 104, at Kirkwood, Del., on the 20th ; lowest temperature, 40, at Sunny Side, on the 8th ; greatest. local monthly range, 21, at Jewell ; mean monthly range, 41.2. Mean maximum temperature, 81.5 ; mean minimum temperature, 62.9.

The temperature was .3 below the normal for the month.

Precipitation (in inches).—Average, 2.22 ; greatest amount, 5.00, at Fenby ; least amount, .72, at Barron Creek Springs.

The precipitation was 1.44 below the normal for the month. Northern-central Maryland had the greatest average precipitation, 2.84.

Wind.—Prevailing direction, northeast. Total movement in miles, Baltimore, 4867 ; Norfolk, Va., 5884 ; Washington. D. C., 3817.

JULY, 1893.

Temperature (degrees).—Monthly mean (for State), 76.5 ; highest monthly mean, 81.3, at Cambridge ; lowest monthly mean, 67.8, at Oakland. Highest temperature, 102, at Denton, on the 26th ; lowest temperature, 43, at Sunny Side, on the 11th and 24th. Greatest local monthly range, 52, at Denton ; least local monthly range, 30, at Cambridge ; mean monthly range, 39.5. Mean maximum temperature, 86.2 ; mean minimum temperature, 66.4.

There was a marked difference in average temperatures, the isotherms of monthly means varying all the way from 81° near Cambridge, Dorchester county, to 68° in the vicinity of Oakland, Garrett county. The extreme western portion of Maryland was, as a matter of course, the coldest section. The next coldest was the northern portion of northern-central Maryland, including portions of Baltimore and Harford counties. Cambridge, Dorchester county, Eastern Maryland, was the station with highest temperature, and the center of the warmest section.

Precipitation (in inches).—Average, 2.97 ; greatest amount, 7.12, at Cambridge ; least amount, 1.38, at Cumberland (1).

The precipitation of the month was distributed very unevenly. The total fall was heavy—considerably above the normal—over a section having a breadth of from 15 to 35 miles, and extending from the mouth of the Delaware River, southwest, into the central portion of Southern Maryland. Large portions of the southern-central and western parts of Maryland, however, received much less than the average amount.

Wind.—Prevailing direction, southwest. Total movement in miles, Baltimore, 4957 ; Norfolk, Va., 5370 ; Washington, D. C., 4133.

AUGUST, 1893.

Temperature (degrees).—Monthly mean (for entire territory covered), 73.5 ; highest monthly mean, 79.2, at Kirkwood, Del.; lowest monthly mean, 63.6., at Sunny Side ; highest temperature, 97, at Boettcherville, on the 25th ; lowest temperature, 44, at Oakland, on the 7th, 8th, 15th, and at Boettcherville, on the 15th. Mean monthly range, 38.1 ; greatest local monthly range, 53, at Boettcherville; least local monthly range, 31, at Solomon's. Mean maximum temperature, 84.4 ; mean minimum temperature, 64.6.

Attention is invited to the contrast between the means of the different sections of the State (see table Monthly Summary of Reports). The mean temperature of Western Maryland, for August, was 69.2, while that of Southern Maryland was 6.5 higher. Northern-Central and Eastern Maryland mean temperatures lie between these extremes, the former section being the cooler, as would be expected.

Precipitation (in inches).—Average, 3.34 ; greatest amount, 6.26, at Fallston ; least amount, 1.61, at Cambridge.

The variation in distribution of the month's rainfall is seen to be very great. Many portions of the State had plenty of rain and some had too much, but complaints of drought were received, also.

Wind.—Prevailing direction, northwest. Total movement in miles, Baltimore, 5222 ; Norfolk, Va., 6092 ; Washington, D. C., 4279.

SEPTEMBER, 1893.

Temperature (degrees).—Monthly mean (for entire territory covered), 65.3 ; highest monthly mean, 71.2, at Cambridge ; lowest monthly mean, 58.0, at Oakland ; highest temperature, 93, at Cambridge, on the 7th ; lowest temperature, 28, at Sunny Side, on the 29th and 30th ; mean monthly range, 48.5 ; greatest local monthly range, 56, at Boettcherville ; least local monthly range, 41, at Receiving Reservoir, D. C. Mean maximum temperature, 74.6; mean minimum, 56.7.

Precipitation (in inches).—Average, 2.76; greatest amount, 6.17, at Millsboro, Delaware; least amount, 1.40, at Sunny Side.

Wind.—Prevailing direction, northwest. Total movement in miles, Baltimore, 5062; Norfolk, Va., 5571; Washington, D. C., 3851.

OCTOBER, 1893.

Temperature (degrees).—Monthly mean (for entire territory covered), 54.9; highest monthly mean, 61.9, at Cambridge; lowest monthly mean, 48.4, at Oakland; highest temperature, 88, at Cambridge, on the 3rd; lowest temperature, 13, at Sunny Side, on the 31st; greatest local monthly range, 68, at Sunny Side; least local monthly range, 47, at Solomon's; mean monthly range, 53.8. Mean maximum temperature, 65.3; mean minimum temperature, 45.6.

Precipitation (in inches).—Average, 4.42; greatest amount, 6.40, at Valley Lee; least amount, 2.59, at Darlington.

Wind.—Prevailing directions, northwest and southeast. Total movement in miles, Baltimore, 5360; Norfolk, Va., 5993; Washington, D. C., 4473.

NOVEMBER, 1893.

Temperature (degrees).—Monthly mean (for entire territory covered), 42.6; highest monthly mean, 48.2, at Cambridge; lowest monthly mean, 34.1, at Sunny Side; highest temperature, 73, at Charlotte Hall, on the 30th, and at Barron Creek Springs, on the 4th; lowest temperature, zero, at Sunny Side, on the 26th. Greatest local monthly range, 66, at Sunny Side; least local monthly range, 37, at Cambridge; mean monthly range, 48.8. Monthly mean maximum temperature, 50.4; monthly mean minimum temperature, 34.5.

Precipitation (in inches).—Average, 3.74; greatest amount, 8.27, at Valley Lee; least amount, 2.01, at Cumberland (2).

Wind.—Prevailing direction, northwest. Total movement in miles, Baltimore, 5422; Norfolk, Va., 6903; Washington, D. C., 4818.

SUMMARY OF THE WEEKLY WEATHER CROP BULLETINS

ISSUED IN 1892 AND 1893.

Week ending April 7, 1892.

In Northern-Central Maryland during the past week the rainfall is reported as having been above the normal, but well distributed, and, with the high temperature and moderate amount of sunshine, generally beneficial to vegetation, particularly wheat.

In Southern Maryland there has been less rainfall, but as much as needed. The warm weather and abundant sunshine is reported as having been favorable to hotbeds, tobacco beds, and crops generally, as well as for farm work.

In Eastern Maryland the total amount of rainfall during the week has been apparently below the normal, fairly well distributed, beneficial to vegetation, and favorable for the advancement of farm work. The temperature has been slightly above the average, apparently, and with the considerable amount of sunshine given, the condition of wheat, oats and grass is much improved.

In Western Maryland the total amount of rainfall during the week has been apparently about the normal, fairly distributed, and beneficial to grain and grass. The temperature for the past seven days has been apparently above the average, and beneficial to vegetation in general. The average amount of sunshine given has been likewise beneficial to all crops.

In Delaware the rainfall during the week is reported as about the normal, the temperature as apparently above the average, and the sunshine average in amount—weather conditions which have proved favorable generally to the advancement of farm work and the growth of crops.

Week ending April 14, 1892.

In Northern-Central Maryland the total amount of rainfall during the week has been below the normal, and fell at the beginning and end of the week. It is reported as being injurious to budding fruit trees, and to fallow ground being prepared for potatoes and early gardening; but as beneficial to wheat. The temperature has been considerably below the average, reaching the freezing point several times during the week. It is not certain, however, that the blossom buds of fruit have been injured. The abundant sunshine of the week counterbalanced, to some extent, the effect of the low temperature.

In Southern Maryland there has not been much rain, and but little need of any. It is supposed that some damage to fruit resulted from the cold wave, and it is reported that tobacco plants have been slightly injured by frosts on the 10th, 11th, 12th, 13th. Tobacco plants and wheat benefited by the abundant sunshine.

In Eastern Maryland there has been a small amount of rain, but it was not generally needed, and has even been injurious to low lands. Wheat, however, is reported as suffering slightly from drouth in the central portion. The temperature has been considerably below the normal and it is supposed that fruit has been injured slightly. The correspondent at Barron Creek Springs reports oats, grass and wheat greatly injured by frost of the 12th, and peas and strawberries much damaged by high winds of the 9th and 12th. Beneficial results from the large amount of sunshine.

The reports from Western Maryland have been too few to permit of a generalization for the whole section.

Week ending April 21, 1892.

In Northern-Central Maryland the rainfall of the week has been above the normal, about twice the average amount falling at Baltimore. It has been fairly well distributed and is reported as being beneficial to wheat, grass, and garden crops in all sections, excepting the southern portions of Harford and Baltimore counties. The week has been cold, its mean temperature at Baltimore being about 10° below the normal, and while no damage from frosts is reported, the growth of crops was much retarded. The amount of sunshine has been below the average, and though of some benefit to wheat, grass, and budding fruit trees, a greater amount was needed.

In Southern Maryland more than the average amount of rain fell. It is reported as follows from the Maryland Agricultural College: Thursday, 14th, .73; Friday, 15th, .20; Monday, 18th, .72; Wednesday, 20th, .15. As seen from the above, it was quite well distributed; but the total amount was greater than necessary, and in some localities it was injurious to wheat and gardens. Benefit is reported from Anne Arundel county. The temperature was considerably below the normal and the amount of sunshine was not large. As a consequence the growth of crops has been generally slow.

In Eastern Maryland the rainfall is reported, from nearly all sections, as in excess of the normal, but fairly distributed ; the temperature as below the average, and the sunshine as small in amount. The above weather conditions have proved generally unfavorable to the growth and development of vegetation, early vegetables, such as potatoes, peas, etc., probably suffering most from the continued low temperature and excess of rainfall. Frosts are reported to have done some damage to peach buds and vegetables on the 16th, 19th and 20th instant.

Week ending April 28, 1892.

In Northern-Central and Western Maryland there was apparently more than the average amount of rainfall, and it is generally considered to have been of benefit to crops, though in some portions of Harford and Carroll counties the opposite opinion prevails. A higher temperature and a greater amount of sunshine were needed. In places the low temperature is reported to have been injurious to potatoes, oats, and other growing crops. The latter portion of the week was all that could be desired.

In Southern Maryland the rainfall for the past week was considerably above the average, and generally injurious except to grass. The temperature for the past seven days was about the normal, though at the beginning it was much below. An average amount of sunshine was given.

In Eastern Maryland the rainfall was apparently above the normal, and, in the northern portions, generally beneficial to wheat, oats, and grass, while reports from the southern portion state that it has been so heavy as to injure all spring crops. All wheat on low lands was more or less injured. The temperature was apparently a little below the average, and there was a small amount of sunshine, to the detriment, in particular, of early vegetables and fruit.

Week ending May 5, 1892.

The past week has been the most favorable of the present season for all sections of Maryland, the weather seeming exactly suited to the various needs of the farmer. The rainfall was below the normal and fell in light showers ; while the temperature and sunshine were considerably above the average. This warm, showery, sunshiny weather proved of the greatest advantage ; in consequence of it, vegetation is much advanced and farm work is progressing finely. Corn planting is under way, or the ground is rapidly preparing for it ; grass and grain have made rapid progress, and apple and cherry trees are in heavy bloom. Peaches have grown to the size of a pea,

and it is stated that the crop on high ground will be large. Truck is
of excellent quality and growing finely. Tobacco plants are thriving
and plenty. In Southern Maryland labor is scarce, and a good class
of working people would be gladly welcomed.

Week ending May 12, 1892.

The weather of the past week, though rather too cool, has been
generally favorable to farming interests throughout the State, and
crops look and promise well. The rainfall was slightly below the
normal, but well distributed. Corn planting is nearly finished.
Wheat is growing rapidly and promises a good crop. Grass is backward in some localities, while in others the growth is good. Tobacco
plants in Southern Maryland are growing finely, but damage by flies
is reported from some places. A hail-storm on the 11th, between
Baltimore and Washington, damaged crops to considerable extent.
The pear crop promises generally well, and early peaches bid fair to
be a good crop in most localities. The outlook for the strawberry
crop is good, and it is stated that peas will be ready to ship from
Eastern Maryland in two or three weeks.

Week ending May 19, 1892.

The conditions of the weather during the past week have been
generally favorable to crops and vegetation. An average amount
of sunshine has been beneficial, but the temperature was rather too
far below the normal. The rainfall was quite general in the early
part of the week, and above the average. In Southern Maryland
some damage to wheat and early vegetables has been reported, and
planting was interrupted. In other sections of the State, however,
wheat has greatly improved and promises well. Grass, though short,
has rapidly advanced. Early tobacco plants are doing well ; but the
late ones have been greatly injured by flies and cold weather. Strawberry fields, in some places, have been infested by insects from
which much damage has resulted, but in others the crop is reported
to be in excellent condition and, with favorable weather, ripe berries
will soon be plentiful. Early vegetables generally are doing well.

Week ending May 26, 1892.

The week was generally too cool, too cloudy, and too wet,
especially in northern sections. Truck grew but slowly, and little
work was done. Wheat, oats and grass look well in most sections
and wheat is beginning to head. Wheat and oats, however, were
somewhat injured by cold and rain in portions of Harford and Anne

Arundel counties, and also in Eastern Maryland. Some wheat has turned yellow for want of warmth and sunshine. Corn-planting is not yet finished in northern sections. Melons injured by wet weather in Southern Maryland and the temperature was too low for the development of peas. Strawberries late. but beginning to ripen. Fruit has been injured by the cold weather, but it is believed that there will be, on the whole, a fair crop. Tobacco plants are doing well and but little injury from the fly is reported. Warmer weather and a greater amount of sunshine needed in all sections and for all crops. Tobacco plants would be benefited by warm showers.

Week ending June 6, 1892.

The warm sunshiny weather has been beneficial to all crops and decidedly favorable for farm work. Wheat looks poor in some of the southern counties, but better in northern and western sections. In Eastern Maryland it has been slightly injured (together with vines and grass) by high winds. Tobacco planting under way in the southern portions of the State and some damage done by "the fly." The hay crop reported excellent in Western Maryland, but short, and in need of rain in other sections. There is the promise of considerable fruit in the northern and western portions of the State, but a poor outlook elsewhere. Truckers are picking and shipping strawberries in Southern Maryland, while in Western Maryland they are in bloom and promise a very large crop. Oats in the west are reported as well started. Showers are needed in all sections, with a continuation of the high temperature and abundant sunshine.

Week ending June 13, 1892.

Crops in the east and the south have suffered slightly from excessive rains, causing corn to turn yellow on low lands and wheat to rust. Tobacco planting reported to be about half finished, and the plants as scarce and backward. The wheat harvest has begun in Eastern Maryland somewhat earlier than last year. In Northern-Central and Western Maryland the weather conditions have been decidedly favorable to growing crops and to the advancement of farm work. Wheat in these sections is growing rapidly, and it is reported that with a continuation of the favorable conditions it will be ready to harvest by the latter part of the month. The weather is favorable for tomato planting. The week closes with the need of sunshine and warm nights.

Week ending June 20, 1892.

The reports of this week indicate that the high temperature which has prevailed, has been of benefit to most sections of the State. In the central and western portions corn is making a rapid growth, haying is progressing favorably, and wheat is rapidly improving. In Eastern Maryland rain is much needed, crops having suffered slightly from drouth. In Southern Maryland corn is making a rapid growth ; tobacco has been slightly injured by cut-worms. The week closes with a need of continued warm weather and an average rainfall for all sections of the State.

Week ending June 27, 1892.

Favorable to the farmer throughout the central and western portions of the State, with the exception of Harford county, where there is a need of rain. Wheat harvesting being actively carried on in Northern Central Maryland with the prospect of an excellent yield. Considerable hay, also, has been cut and saved in good condition. In the east and south, grass, tobacco and garden vegetables have suffered from drouth. The wheat and hay harvest in Eastern Maryland is finished and the crops have been saved in good condition. A continuance of the average temperature and sunshine, with more rain in Eastern and Southern Maryland, is needed.

Week ending July 4, 1892.

On July 1st heavy storms, accompanied by hail and high winds, did considerable damage to crops and vegetables in northern-central, eastern and southern parts of the State.

On July 4th, reports stated that harvesting was approaching completion and that threshing had commenced. It was estimated that the wheat yield would be about the average. Early planted tobacco was promising ; fruit in eastern sections had been damaged by high winds ; corn, grass and clover were in a flourishing condition.

Week ending July 11, 1892.

Wheat harvesting about finished, except in Western Maryland. The outlook for the corn crop, upon the whole, good. Reports of the tobacco crop rather unfavorable. Oats poor, except in Western Maryland, where the outlook good for an average crop. Hay reported heavy in western sections, and fair elsewhere. Early potatoes excellent. The fruit crop light and few peaches being shipped.

Week ending July 18, 1892.

The wheat harvest over, except in Western Maryland. Threshing in progress, and nearly finished in eastern and southern sections. A full average yield reported from most localities. Corn in Southern Maryland somewhat injured by rain, and in other portions of the State by drought and worms; but the general outlook for the crop is good. Hay shorter than anticipated. All vegetation needs rain. Shipments of potatoes and fruit light.

Week ending July 25, 1892.

Crops suffering slightly from drought, though tobacco reported to have improved during the preceding week. The weather favorable for haying, threshing, and outdoor work generally. There is a general scarcity of fruit, and the prospects for the corn crop continue good.

Week ending August 1, 1892.

Potatoes being dug and the yield reported light. Corn, oats, millet, tomatoes, clover, pasture and fallow improved, and tobacco growing rapidly. It is thought that a full crop of buckwheat will be secured in Western Maryland.

Week ending August 8, 1892.

Condition of all crops improved. More rain needed in Western Maryland and in portions of Southern Maryland. The peach crop reported as fair in some localities, and tomatoes beginning to ripen. Fallow plowing in progress.

Week ending August 15, 1892.

All growing crops more or less injured by drought. The weather favorable for harvesting oats in Western Maryland. The crop an average one.

Week ending August 22, 1892.

Drought has injured all crops. Fears entertained that the corn yield will be much below the average. Wheat-threshing in progress in Western Maryland, the yield being reported as from 20 to 30 bushels per acre on good land. Fruit reported scarce.

Week ending August 29, 1892.

Late corn, tobacco, late potatoes, and gardens have suffered a great deal from drought, and it is thought that all crops will be short. Farmers in some portions of Southern Maryland have begun cutting their corn and tobacco. The ground is so hard that fallowing is with difficulty continued. In some sections it has been discontinued altogether. The best results have been obtained along the northern border of Maryland, including the western portion, where light showers have fallen. Farmers saving fodder in Eastern Maryland. The month closes with the need of rain in all sections.

Week ending September 5, 1892.

Reports from Northern-Central Maryland state that the farmers are cutting corn and preparing the ground for seeding. The yield of corn will be below the average. Rain is much needed.

Reports from Southern Maryland state that all crops are suffer. ing from drouth and that the yields will be short in consequence thereof. Rain, to be of benefit, must come soon.

Reports from Western Maryland state that buckwheat, corn, potatoes, and all other full crops are greatly suffering from drouth. The weather is favorable for farm work, with the exception of wheat seeding. Light frosts occurred in Garrett county on the 2d and 3d.

Reports from Eastern Maryland state that the weather has been favorable for the saving of fodder. Clover, grass, late corn, potatoes, and tomatoes are reported as suffering from the drouth, though the scattered showers of Monday and Wednesday did a great deal of good.

Week ending September 12, 1892.

In Northern-Central Maryland farmers are getting ready for seeding, where it is possible to plow. In the western portion the heavy showers of Thursday much improved the condition of the ground and benefited late crops. In the eastern portion the drought continues, plowing is almost impossible, and pastures and fall crops are suffering. Corn cutting is in progress and on many farms an average crop will be secured.

In Southern Maryland all crops greatly damaged by the drought which still continues, and the early crops are being secured in order to prevent waste. Seeding delayed by the dry weather.

In Western Maryland buckwheat, corn, and garden truck slightly injured by frosts on the 6th and 7th. All vegetation greatly suffering from drought. Seeding will be late unless rains come soon. Wheat threshing in progress; grain turning out well. Apples reported better than anticipated, peaches as excellent, and garden truck of fair quality though below the average in amount.

In Eastern Maryland the weather has been very favorable for fodder saving and general farm work, but the fallow ground is too dry for proper working, and pastures, corn, potatoes, and all growing crops are very much in need of rain.

Week ending September 19. 1892.

In Northern-Central Maryland pastures and fall crops saved by the timely rains. Late fall plowing and seeding are now progressing rapidly. Heavy rains slightly damaged standing corn and recently seeded fields.

Southern Maryland:—Corn and tobacco cutting in progress. Recent rains have greatly improved late crops. Farmers are preparing for wheat seeding.

In Western Maryland plowing for wheat has been made less difficult by the recent rains. More rain is needed. Apples suffered from the high winds of the 13th and 14th. Buckwheat threshing in progress, and the yield is reported as from 10 to 12 bushels per acre on good land. No frosts occurred during the week.

In Eastern Maryland the weather has been favorable for all kinds of farm work, but more rain is needed for plowing and for growing crops. The fodder is about all saved in good condition, and corn is being cut. Standing corn was slightly injured by the high winds on Tuesday.

Week ending September 26, 1892.

In Northern-Central Maryland wheat and rye seeding progressing, though somewhat delayed by rains. Pasture much improved. Bright weather now needed for maturing corn. In many localities corn cutting has begun, and an average crop is reported. Tomato canning going on.

In Southern Maryland tobacco, corn, and pastures have been greatly improved by the late rains, and grain seeding is progressing favorably and rapidly. The abundant sunshine has been very beneficial to maturing tobacco and corn, and the equable temperature has been favorable to tobacco, both in the house and in the field. The corn harvest is going on, and the crop is reported as generally excellent.

In Western Maryland weather favorable for the farmer. Wheat seeding in progress and corn cutting well advanced. Peach shipments are reported as heavy.

In Eastern Maryland grapes, late tomatoes, and vegetables benefited by the moderate rains of the week. More rain is needed in some localities, to put ground in proper condition for seeding, while in others fair weather would be most beneficial. The corn crop is reported as below the average.

Week ending October 3, 1892.

In Northern-Central Maryland the tomato season is about over. Corn is all cut, and the shocks are said to average fewer and smaller than usual. Pasturage is failing fast. Fall seeding mostly done, and some winter wheat is already up. Light frosts occurred, but no damage has, as yet, been reported. Seeding almost finished.

In Southern Maryland tobacco nearly all housed ; crop is reported as very short in quantity, above average in color, and all sound. Potatoes are a dismal failure. Farmers will be late seeding, as the ground is still hard and dry. Pastures have considerably improved.

In Western Maryland the early sown wheat, though somewhat retarded by the drought, is growing nicely. Corn is most all in shock, and buckwheat threshing in progress. The crop is short on account of dry weather. Fruit and vegetables plenty.

In Eastern Maryland all crops were much benefited by a good rain Sunday morning, which was the heaviest that has fallen in this section during the season. This rain put the ground in excellent condition for seeding, which is now in progress. Potatoes almost a failure in some sections of Delaware.

Week ending April 10, 1893.

In Western Maryland the threatened drouth was relieved by copious rains at close of week. The rain was much needed, as late sown wheat was looking bad in some places. Oats sowing and potato planting in progress. Grass looking well. Prospects for fruit favorable.

In Northern-Central Maryland a great deal of spring work accomplished ; potatoes planted ; oats sown ; soil in excellent condition ; growing grain rather better than average ; fruit promising well ; peach buds uninjured.

In Southern Maryland farm work far advanced ; the rain that fell was of great benefit to wheat, oats, grass, and all crops ; some clover damaged by dry weather ; tobacco plants growing well. Showers of the 7th enabled the setting out of early cabbage and strawberry plants ; the cloudy weather following was favorable. Prospect for peaches, favorable.

In Eastern Maryland prospect reported good for full peach crop, peach buds being in good condition, abundant, and swelled nearly to blooming. Fruit, generally, not so early as usual, owing to severe winter. Wheat and grass have been backward, but will greatly improve owing to the recent rains. Weather has been very favorable for all kinds of farm work.

Week ending April 17, 1893.

In Western Maryland rains of past week beneficial to growing crops, but unfavorable for farm work. It is thought that the cold weather of Saturday may have injured the fruit. Warm weather needed.

In Northern-Central Maryland outdoor work, to some extent, interrupted by the well distributed and much needed rain. All crops improved; clover short; wheat fields looking well, with few exceptions, but warmer weather with sunshine needed. Preparations for corn planting well advanced. Ground in excellent condition. Fruit prospects good.

In Southern Maryland plowing backward; oats coming up; clover and grass seed backward in sprouting; tobacco beds late; peas growing nicely; cherry, peach, and pear trees beginning to bloom. Full crops of all fruits anticipated; some fears, however, as to results of Saturday night's cold weather.

In Eastern Maryland the liberal rains of the week, well distributed, and slightly in excess of the normal, greatly benefited all crops in all sections. Wheat and grass growing rapidly. Peach trees beginning to bloom, and an abundant crop promised. Other fruit ready to blossom. Pear crop bids fair to be a full one. Corn ground nearly all broken. Farm work well advanced.

Week ending April 24, 1893.

In Western Maryland spring plowing well advanced and corn will be planted during ensuing week. A delay of farm work on account of the rain is reported. It is believed that there has resulted no injury to fruit crop on account of the frost. Wheat has improved.

In Northern-Central Maryland wheat, generally, looks well; the weather, however, has been too wet and cold for all growing crops. Peaches in some places thought to have been injured by frost of 24th. Plowing delayed by wet weather.

In Southern Maryland wheat not yet recovered from past drouth. Oats and tobacco plants growing nicely. Some apple blossoms show effects of frost of 16th, though fruit, generally, is not thought to be injured. Corn planting has begun. The rains of the week have been beneficial, but warmer weather and more sunshine is needed.

In Eastern Maryland, the past week, on account of the heavy rains and low temperature, has not been favorable, generally, to the rapid preparation of corn ground, though better reports have come in from some localities. Some corn has been planted. Wheat is stated to have wintered well and to be in good condition. Grass somewhat backward, but now improving. Oats beginning to look green. Cherry and pear trees in blossom. Peach trees in bloom, and not thought to have been injured by the cold weather.

Week ending May 1, 1893.

In Western Maryland wheat fair; grass short; oats growing slowly; peach trees in bloom—reported not injured by frost.

In Northern-Central Maryland wheat, oats, and grass looking fairly well; farm work delayed, and corn planting late on account of wet weather; fruit probably not injured by frost.

In Southern Maryland grass and grain not up to average, but benefited by the copious rains. Oats growing nicely; plowing retarded; tobacco plants plenty and doing well; fruit trees in bloom, and thought uninjured by frosts; early peas looking well; beans and corn, planted over two weeks ago, now coming up; some tomato plants set out.

In Eastern Maryland weather unfavorable for farm work; little corn planting yet done; peach trees, and fruit trees generally, in bloom and reported uninjured by cold weather; strawberries in blossom and injured slightly by frosts: raspberries looking well; wheat and grass in good condition.

Week ending May 8, 1893.

Wheat, oats and rye growing well. Good growth of grass reported in some sections, but poor in others. Very little corn planted. Potatoes coming up. Tobacco plants thriving. Fruit prospects good.

Week ending May 15, 1893.

Excellent week for farm work. Wheat, oats, rye and potatoes getting a good growth. A great deal of corn planted, and early planted growing nicely. Grass excellent in some sections, but poor in others. Tobacco plants plentiful. Strawberries growing well. Excellent prospects of large crops of peaches and apples.

Week ending May 22, 1893.

Farm work somewhat retarded by rains. Wheat, rye and oats show a healthy growth generally. Grass reported short in some localities. A large acreage of corn being planted; some coming up. Tobacco plants abundant. Planting commenced and large acreage anticipated. Strawberries ripening. Good prospects for fruit of all kinds, especially peaches.

Week ending May 30, 1893.

Wheat heading; large harvest anticipated within a month. Oats growing well. Corn coming up nicely; some damage from cut-worms. Grass improving, but in some sections backward, and pasture short. Tomato plants scarce. Potatoes coming up. Gardens improving. Peas will be ready for market during the week. Vegetables promise well. Tobacco planting in progress, with plenty of plants. Strawberries ripening. Peaches and apples will be abundant.

Week ending June 5, 1893.

The wheat crop promises a large yield. Corn getting a healthy growth. Grass doing very well, except in Eastern Maryland. Fair prospects for the apple crop in Western Maryland. The peach crop promises well. Strawberries ripening and bringing fair prices. Large acreage of tobacco anticipated, owing to advance in prices; plants being set out.

Week ending June 12, 1893.

General appearance of all crops encouraging. Barley about ready for harvest. Large fruit dropping to some extent, but the prospects remain good. Truck in excellent condition.

Week ending June 19, 1893.

Wheat harvest begun and large yield expected. Oats heading. Corn and potatoes looking well. Hay-making in progress. Tobacco planting being pushed rapidly. Fruit crop fair.

• *Week ending June* 26, 1893.

Wheat harvesting progressing rapidly; an excellent yield, as a rule. Haying in progress, with fair results. Corn looking well. Buckwheat sowing being pushed. Tobacco improved by rains. Vegetables doing well. Fruit prospects fair.

Week ending July 4, 1893.

Wheat harvest about half over, and a good yield is expected. Oats look the best for years. Peaches are growing finely, and a heavy market should result. Apples promise to give a fair crop in most sections. Peas are being shipped. A fair crop of corn may be looked for. The clover crop will be good, but the prospects for timothy are not so favorable.

Week ending July 10, 1893.

Buckwheat progressing nicely. Haying commenced. Raspberries short, but the plum and peach crops are good. Cherries will give an average yield. Gardening prospects fine ; potatoes and tomatoes doing especially well. Oats are heading nicely. Tobacco improving, but suffering from cut-worm in some sections. Fruit prospects good. Cucumbers a little backward.

Week ending July 17, 1893.

Hay-making about over, and fair yield. Potato crop above average. Some oats being cut. Corn making splendid growth. Second crop of clover coming in. Vegetables plentiful. Tobacco growing nicely. Apple crop best in years. Wheat threshing nearly completed, and yield above expectations. Potato vines growing vigorously, and indications point to a big crop. Peaches are being shipped in large quantities, and the quality is good.

Week ending July 24, 1893.

Corn coming into tassel, and the crop looks well. Wild berries a failure. Plowing for fall crops has commenced. Potatoes ripening. Fruit suffering some from insects. Gardens look dry. Early peach and apple crops good. Reports from Eastern Maryland state that the yield of wheat is the largest in years ; the quality, too, is excellent.

Week ending July 31, 1893.

Corn injured by drought, but still looks well in most localities. Some potatoes dug, and average yield reported. Vegetation, generally, suffering for want of rain. Some plowing for fall seeding being done. Gardens and pastures would be much benefited by rains. Heavy yields of peaches reported from some orchards. Prospect for apples is good. Satisfactory yield of wheat, generally, but light in some places.

Week ending August 7, 1893.

Prospects for corn and late potatoes poor in western sections. Fruit yield large and of good quality. In portions of Northern-Central Maryland vegetation suffered slightly from the want of rain. Fallowing began in Southern Maryland.

Week ending August 14, 1893.

All crops in Western Maryland injured by drought, except buckwheat. Tomato crop very light. Corn and late tobacco slightly improved in some sections of Anne Arundel county. In Eastern Maryland threshing nearly over and yield average.

Week ending August 21, 1893.

Drought continued in many sections, seriously injuring crops and pastures. It was broken in portions of Northern-Central and Eastern Maryland. to the great improvement of vegetation.

Week ending August 28, 1893.

Drought broken in most sections, with a general improvement in growing crops. Some areas still in need of rain. Ground being prepared for winter wheat. Peaches ripening in Western Maryland. Peach crop large in Southern Maryland. Fair crop of early tobacco being housed.

Week ending September 4, 1893.

Growing crops greatly improved by soaking rains, but high winds damaged fruit, fruit-trees, corn-fodder, tomatoes, buckwheat and tobacco; considerable wheat sown; ground in excellent condition : grass improved : corn cutting begun; fruit plenty : tomato-canning in progress.

Week ending September 11, 1893.

Considerable tobacco housed, late benefited by rain; fair yield of buckwheat; corn and fodder damaged by high winds, cutting commenced; some fall seeding done; peach season approaching close in eastern and southern sections : late peaches in western section improved by rain.

Week ending September 18, 1893.

Ground is in excellent condition for seeding, which is in progress ; housed tobacco slightly injured by warm, wet weather early in week ; pastures and vegetables greatly improved ; corn cutting and fodder saving progressing rapidly : peaches nearly gone.

Week ending September 25, 1893.

Good weather for plowing, corn cutting and wheat seeding : fodder and tobacco now in progress; pastures and late tobacco greatly improved ; peaches and apples excellent in western sections : large quantities of tomatoes being picked in eastern portion.

Week ending October 2, 1893.

In Western Maryland wheat seeding about finished ; early sown growing rapidly. Corn all cut and husking in progress : yield not up to expectations. Buckwheat threshing nearly finished; yield somewhat less than average. Potatoes fair; digging about completed.

In Northern-Central Maryland wheat seeding approaching completion, with ground in excellent condition. Pastures improved. Corn nearly all cut; average crop. Potato crop better than anticipated. Cabbage improved. Some fields of buckwheat may have been injured by frost. Fruit falling off; remaining apples being picked.

In Southern Maryland weather conditions excellent for all farm work. Late tobacco has grown rapidly and matured well; much of the crop still out, but farmers hurrying to save it. Ground in excellent condition for wheat seeding, which is in progress. Corn hard to cut, but not much damaged; short in some places. Potatoes few. Apple crop generally fair.

In Eastern Maryland farmers very busy saving corn crop. Some seeding done. More rye will be sown than usual. Tomato picking still in progress. Apple crop fair.

MONTHLY SUMMARY OF REPORTS FOR JANUARY, 1892.

Stations.	Counties.	Altitude above sea in ft.	Latitude.	Longitude.	Monthly Mean.	Mean of Max.	Mean of Min.	Degrees, Max.	Degrees, Min.	Monthly Range.	Total Precipitation.	Clear Days.	Fair Days.	Cloudy Days.	Rainy Days. (.01 inch or more.)	Prevailing Wind.
WESTERN MARYLAND.																
Boettcherville	Allegany		39°33'	78°48'	*31			58	0	58	4.70				9	
Cumberland (1)	Allegany	700	39 39	78 45	35	42	27	57	8	49	3.12	5	9	17	8	N.W.
Cumberland (2)	Allegany	700	39 39	78 45	30	35	25	52	5	47	3.18	11	6	14	7	
Hagerstown	Washington					39	24	56	8	48	4.06				9	N.W.
NORTHERN-CENTRAL MD.																
Baltimore		179	39 17	76 37	32	38	26	58	12	46	6.42	10	10	11	12	N.W.
Darlington	Harford	300	39 47	76 14	28	37	21	58	7	51	5.63	16	0	15	7	N W.
Dist. Res., D. C.			39 9	77 0	†32			59	5	54	5.33				15	
Fallston	Harford	300	39 30	76 24	*30			67	11	46	5.54				11	
Frederick	Frederick	400	39 24	77 24	31	40	23	57	5	52	5.48				12	
Great Falls	Montgomery		39 0	77 14	†31			60	5	55	4.68				12	
McDonogh	Baltimore	545	39 23	76 46	30	38	22	64	8	56	4.95				10	
Mt. St. Mary's	Frederick	720	39 43	77 20	30	37	20	56	8	48	5.32	12	3	16	9	N.W.
New Market	Frederick	500	39 10	77 15	*29			49	7	42	4.46	10	10	11	11	N.W.
Rec. Res., D. C.			38 52	77 0	†32			58	9	49	5.29				14	
Taneytown	Carroll		39 40	77 9							1.*7				9	
Washington, D. C.		112	38 52	77 0	32	39	24	62	4	58	5.84	11	5	15	12	N.W.
Woodstock College	Baltimore	400	39 19	76 51	31	37	21	61	0	61	4.58	4	11	16	12	N.W.
SOUTHERN MARYLAND.																
Bryantown	Charles				32	42	23	66	5	81						
Charlotte Hall	St. Mary's		38 28	76 48	31	39	22	66	3	63	4.03					
Jewell	Anne Arundel		38 44	76 36	†28			57	4	53	4.33			1	12	
Leonardtown	St. Mary's		38 18	76 40												
Solomon's	Calvert	20	38 19	76 27	33	36	28	66	10	56	5.09	5	9	17	12	N.W.
EASTERN MD. AND DELAWARE.																
Barron Ck. Springs	Wicomico	25	38 30	75 39	35	44	28	70	10	60	4.06	6	9	16	8	N.W.
Denton	Caroline															
Dover, Del	Kent	40	39 10	75 30	33	41	25	68	8	58	4.89	11	7	13	11	N.W.
Easton	Talbot	35	38 42	76 6	34	41	26	62	10	52	4.39	12	7	12	11	N.W.
Kirkwood, Del	New Castle		39 35	75 41	†31			70	10	60						
Seaford, Del	Sussex		38 40	75 35	34	43	24	56	7	49	4.78				10	
‡VIRGINIA.																
Norfolk		43	36 51	76 17	41	48	23	72	19	53	4.99	10	9	12	13	N.
AVERAGES	Western Maryland				32.0	38.7	25.3			50.5	3.76	8.0	7.5	15.0	7.5	N.W.
	Northern-Cent'l Md				30.7	38.0	22.4			51.5	5.03	10.5	6.5	14.0	11.1	N.W.
	Southern Maryland				31.0	39.0	24.3			58.2	4.48	5.0	9.0	17.0	12.0	N.W.
	Eastern Md. and Del				33.4	42.2	25.8			55.8	4.54	9.7	7.7	13.7	10.0	N.W.
	Entire territory				31.8	39.5	24.4			54.0	4.45	8.3	10.2	14.9	10.2	N.W.

NOTE.—Letters of the alphabet are used to indicate the number of days that are missing from record: e.g a—one day, b—two days, etc. (1)—H. SHRIVER. (2)—E. T. SHRIVER. * From tri-daily readings. † From bi-daily readings. ‡ Omitted in computing averages.

MONTHLY SUMMARY OF REPORTS FOR FEBRUARY, 1892.

STATIONS.	COUNTIES.	Altitude above sea in ft.	Latitude.	Longitude.	Monthly Mean.	Mean of Max.	Mean of Min.	Degrees, Max.	Degrees, Min.	Monthly Range.	Total Precipitation.	Clear Days.	Fair days.	Cloudy Days.	Rainy Days. (.01 inch or more.)	Prevailing Wind.
WESTERN MARYLAND.																
Boettcherville.	Allegany		39°33′	78°48′	*37			60	-8	68	1.95				6	
Cumberland (1).	Allegany	700	39 39	78 45	38	45	31	64	3	61	1.62	9	2	18	7	N. W.
Cumberland (2).	Allegany	700	39 39	78 45	34	40	29	54	2	58	1.72	5	7	15	7	
Hagerstown	Washington				36	44	28	60	-2	62	1.57	8	6	16	4	
NORTHERN-CENTRAL MD.																
Baltimore		179	39 17	76 37	37	43	31	57	14	43	2.41	7	10	12	14	N E.
Darlington	Harford	300	39 47	76 14	28	42	28	55	12	43	1.82	15	0	14	9	N.W.
Dist. Res., D. C.		39	9	77 0	†36			54	4	50	1.77				7	
Fallston	Harford	300	39 30	76 24	*34	40	30	53	12	41	2.54				9	N. E.
Frederick	Frederick	400	39 24	77 24	36	44	28	56	1	56	2.69				10	
Great Falls	Montgomery		39 0	77 14	†36			55	4	51	1.56				9	
McDonogh	Baltimore	545	39 23	76 46	34	41	28	52	10	42	2.01				11	
Mt. St. Mary's	Frederick	720	39 43	77 20	32	40	24	52	10	42	2.63	7	7	15	9	N.W.
New Market	Frederick	500	39 10	77 15	*35			53	10	43	2.92	9	3	17	8	N.
Rec. Res., D. C.			38 52	77 0	†36			54	5	49	1.75				7	
Taneytown	Carroll		39 40	77 9							2.50				7	
Washington, D. C..		112	38 52	77 0	37	44	30	61	6	55	3.64	7	0	13	12	N. E.
Woodstock College.	Baltimore	400	39 19	76 51	35	42	27	53	-3	56	2 45	9	5	15	7	N.W.
SOUTHERN MARYLAND.																
Jewell	Anne Arundel		38 44	76 36	†34			46	8	38	2.75	9	7	13	7	
Leonardtown	St. Mary's		38 16	76 40	37	43	31	54	16	38	2.95	9	4	16	8	N. E.
Solomon's	Calvert	20	38 19	76 27	34	45	29	55	10	45	2.50	6	6	17	10	N. E.
EASTERN MD. AND DELAWARE.																
Barron Ck. Springs.	Wicomico	25	38 30	75 39	36	43	29	50	9	51	3.28	10	5	14	11	N.W.
Denton	Caroline										2.48					
Dover, Del	Kent	40	39 10	75 30	36	43	29	58	12	44	3.02	10	4	15	11	N. E.
Easton	Talbot	35	38 42	76 6	38	45	30	60	13	47	2.98	10	3	16	10	N. E.
Kirkwood, Del	New Castle		39 35	75 41	†35			52	12	40						
Seaford. Del	Sussex		38 40	75 35	36	45	28	59	8	51	2.02				10	
‡ VIRGINIA.																
Norfolk		43	36 51	76 17	41	49	35	65	19	46	5.32	10	7	12	12	N.
AVERAGES {	Western Maryland				36.2	43.0	29.3		61.8	1.72	8.1	5.0	16.0	6.0	N. W.	
	Northern-Cent'l Md				34.7	42.0	28.2		47.5	2.36	9.0	5.7	14.3	9.2	N. E. N.W.	
	Southern Maryland				34.7	44.0	30.0		40.3	2.73	8.0	5.7	15.3	8.3	N. E.	
	Eastern Md. and Del				36.2	44.0	29.0		46.6	2.75	10.0	4.0	15.0	10.5	N. E.	
	Entire territory				35.4	43.2	29.1		49.0	2.39	8.8	5.8	15.2	8.5	N. E.	

NOTE.—Letters of the alphabet are used to indicate the number of days that are missing from record:
e. g. a—one day, b—two days, etc. (1)—H. SHRIVER. (2)—E. T. SHRIVER. * From tri-daily readings.
† From bi-daily readings. ‡ Omitted in computing averages

MONTHLY SUMMARY OF REPORTS FOR MARCH, 1892.

STATIONS.	COUNTIES.	Altitude above sea in ft.	Latitude.	Longitude.	TEMPERATURE. Monthly Mean.	Mean of Max.	Mean of Min.	Degrees, Max.	Degrees, Min.	Monthly Range.	Total Precipitation.	Clear Days.	Fair Days.	Cloudy Days.	Rainy Days. (.01 inch or more.)	Prevailing Wind.
WESTERN MARYLAND.																
Boettcherville	Allegany		39°33'	73°48'	*39			60	16		44 2.50				9	
Cumberland	Allegany	700	39 33	78 45	40	48	33	67	19	48	2.63	10	2	19	7	
Cumberland	Allegany	700	39 39	78 45	36	42	30	60	16	44	2.62	10	8	13	10	
Hagerstown	Washington				36	46	27	68	16	50	4.80				8	
NORTHERN-CENTRAL MD.																
Baltimore		179	39 17	76 37	37	44	31	66	20	45	7.20	13	7	11	13	N.W.
Darlington	Harford	390	39 47	76 14	35	42	27	58	13	45	3.63	15	3	13	10	N.W.
Dist. Res., D. C.			39 9	77 0	†37			60	19	41	7.04				19	
Fallston	Harford	300	39 39	76 24	*34			55	16	39	5.86				11	
Frederick	Frederick	400	39 24	77 24	37	44	29	57	11	46	5.58				14	
Great Falls	Montgomery		39 0	77 14	†37			58	19	38	5.58				13	
McDonogh	Baltimore	545	39 23	76 46	35	42	28	58	14	44	5.87				13	
Mt. St. Mary's	Frederick	720	39 43	77 20	34	41	28	58	13	45	6.67	12	7	12	12	N.W.
New Market	Frederick	500	39 10	77 15	*36			60	16	44	5.22	11	7	13	12	N.W.
Rec Res., D. C.			38 52	77 0	†37			58	19	39	5.27				13	
Taneytown	Carroll		39 40	77 9							5.93				12	
Washington, D. C.		112	38 52	77 0	38	45	30	66	17	49	5.70	12	7	12	16	N.W.
Woodstock College.	Baltimore	400	39 19	76 51	38	45	28	62	11	51	4.27	8	12	13	14	N.W.
SOUTHERN MARYLAND.																
Charlotte Hall	St. Mary's		36 28	76 48	38	47	28	58	0	58	3.75		11	18	9	
Jewell	Anne Arundel		38 44	76 36	†36			54	18	36	6.35	16	3	12	13	
Leonardtown	St. Mary's		38 18	76 40	36	44	51	60	21	39	4.09	11	14	6	12	N.W.
Solomon's	Calvert	20	38 19	76 27	36	48	27	62	19	43	4.58	8	4	9	11	N.W.
EASTERN MD. AND DELAWARE.																
Barron Ck. Springs.	Wicomico	25	38 30	75 39	38	45	30	66	17	49	4.85	7	11	12	15	N.W.
Denton	Caroline				36			51	20	31	5.60		12	19	6	E.
Dover, Del.	Kent	40	39 10	75 30	37	45	30	62	18	44	4.59	14	4	13	15	N.W.
Easton	Talbot	36	38 42	78 6	39	48	30	61	18	43	4.98	10	9	12	15	N.W.
Kirkwood, Del.	New Castle		39 35	75 41	†36			55	12	43					12	
Seaford, Del	Sussex		38 40	75 35	37	48	29	65	19	46	6.73				13	
‡VIRGINIA.																
Norfolk		43	36 51	76 17	44	51	36	72	24	48	3.61	10	9	12	14	N. E.
AVERAGES { Western Maryland					37.8	45.3	30.0			46.5	3.14 10.0	5.0	16.0	8.8		
Northern-Cent'l Md.					36.0	43.3	28.4			43.9	5.76 11.5	7.2	12.3	12.1	N.W.	
Southern Maryland..					37.0	46.3	28.7			44.0	5.22 11.7	5.5	11.2	11.2	N.W.	
Eastern Maryland					38.3	46.5	29.8			42.7	5.35 10.3	9.0	11.5	12.8	N.W.	
Entire territory					37.3	45.4	29.2			44.3	4.87 10.9	6.7	12.5	11.2	N.W.	

NOTE.—Letters of the alphabet are used to indicate the number of days that are missing from record: e. g. a—one day, b—two days, etc. (1)—H. SHRIVER. (2)—E. T. SHRIVER. *From tri-daily readings. † From bi-daily readings. ‡ Omitted in computing averages.

MONTHLY SUMMARY OF REPORTS FOR APRIL, 1892.

STATIONS.	COUNTIES.	Altitude above sea in ft.	Latitude.	Longitude.	TEMPERATURE.						Total Precipitation.	Clear Days.	Fair Days.	Cloudy Days.	Rainy Days. (.01 inch or more.)	Prevailing Wind.
					Monthly Mean.	Mean of Max.	Mean of Min.	Degrees, Max.	Degrees, Min.	Monthly Range.						
WESTERN MARYLAND.																
Boettcherville......	Allegany.....		39°39'	78°48'	*48.8	58.1	39.4	86	23	63	3.50	...	...	...	13	.....
Cumberland (1)....	Allegany..	660	39 39	78 46	52.8	61.5	44.1	87	30	51	3.51	10	3	17	9	W.
Cumberland (2). .	Allegany.....	700	39 39	78 45	49.3	58.7	41.0	82	26	58	3.21	8	12	10	8	.....
Rogemont	Washington ..		39 45	77 29	49.5	58.9	40.1	86	24	62		...	...	...		N.W.
NORTHERN CENTRAL MD.																
Baltimore........		179	39 17	76 36	51.6	58.9	44.2	83	32	51	3.15	11	8	11	12	N.W.
Darlington........	Harford......	300	39 39	76 14	49.4	58.8	40.1	80	28	52	2.05	18	3	9	8	N.W.
Dist. Res., D. C....			38 52	77 0	†61.9			79	30	46	5.28	...	...	...	13	
Fallston........	Harford......	450	39 51	76 24	*48.8			78	22	50	2.80	...	...	...		
Frederick........	Frederick....	280	39 24	77 18	50.4	59.5	41.3	83	30	50	2.36	14	5	11	7	.....
Great Falls......	Montgomery..		39 0	77 14	†51.8			80	34	46	4.30	...	...	...	13	
McDonogh........	Baltimore....	535	39 23	76 44	49.4	57.0	41.9	74	30	44	2.03	...	...	...	10	
Mt. St. Mary's....	Frederick....	720	39 41	77 21	49.0	57.4	40.7	81	29	52	2.80	11	16	8	12	N.W.
New Market......	Frederick....	500	39 23	77 18	*50.0			82	31	49	3.05	9	7	14	8	S. W.
Rec. Res., D. C....			38 52	77 0	†51.8			79	33	44	5.34	...	...	...	14	
Taneytown........	Carroll......		39 40	77 9							2.96	...	...	...	10	
Washington, D C..		112	38 52	77 0	51.4	59.7	43.2	81	31	60	4.52	12	7	11	14	N.W.
Woodstock College.	Baltimore...	382	39 20	76 49	50.4	60.4	40.5	80	25	65	3.02	8	10	12	10	N.W.
SOUTHERN MARYLAND.																
Charlotte Hall....	St. Mary's...		38 28	76 48	d52.0	62.4	41.7	85	22	37	4.47	...	...	...	8	.....
Leonardtown.....	St. Mary's...		38 18	76 40	51.8	58.6	44.9	81	30	51	5.08	18	1	11	8	N.W.
Solomon's........	Calvert...	20	38 19	76 27	51.0	61.0	41.0	79	32	47	5.23	5	7	18	9	N.W.
EASTERN MD. AND DELAWARE.																
Barron Ck. Springs.	Wicomico....	25	38 30	75 39	51.3	59.4	43.3	79	29	50	6.66	10	10	10	9	S. E.
Dover, Del........	Kent...		39 9	75 31	51.9	61.0	42.8	81	30	51	4.06	16	6	8	9	N.W.
Easton........	Talbot.....	35	38 42	76 6	54.0	64.0	44.1	82	30	52	4.51	16	4	10	11	N.W.
Kirkwood, Del....	New Castle...		39 35	75 40	†48.0			76	30	48		...	...	...		
Seaford, Del......	Sussex......		38 40	75 35	53.0	63.5	42.8	81	29	52	5.58	...	...	...	9	
‡VIRGINIA.																
Norfolk....		43	36 51	76 17	56.0	64.2	47.9	83	33	50	6.86	13	7	10	12	N. E.
AVERAGES {	Western Maryland..				50.4	58.8	41.4			59.5	3.47	9.0	7.5	13.5	10.0	W. N.W.
	Northern-Cent'l Md.				50.5	58.8	41.7			49.9	3.37	11.9	8.0	10.0	10.9	N.W.
	Southern Maryland.				51.6	60.7	42.5			51.8	4.93	11.5	4.0	14.5	8.5	N.W.
	Eastern Md. and Del.				51.6	62.0	43.2			50.2	5.21	14.0	6.7	9.3	9.5	N.W.
	Entire territory.....				50.0	59.8	42.1			51.8	3.89	11.9	7.1	11.1	10.2	N.W.

NOTE.—Letters of the alphabet are used to indicate the number of days that are missing from record: e. g. a—one day, b—two days, etc. (1)—H. SHRIVER. (2)—E. T. SHRIVER. * From tri-daily readings. † From bi-daily readings. ‡ Omitted in computing averages.

MONTHLY SUMMARY OF REPORTS FOR MAY, 1892.

STATIONS.	COUNTIES.	Altitude above sea in ft.	TEMPERATURE.							Total Precipitation.	Clear Days.	Fair Days.	Cloudy Days.	Rainy Days. (.01 inch or more.)	Prevailing Wind.
			Latitude.	Longitude.	Monthly Mean.	Mean of Max.	Mean of Min.	Degrees, Max.	Degrees, Min.	Monthly Range.						
WESTERN MARYLAND.																
Boettcherville	Allegany	....	39°39'	78°48'	*61.8	71.4	52.3	*6 35		51	4.70	....	....	....	16	....
Cumberland (2)	Allegany	700	39 39	78 45	62.2	70.7	53.9	84 41		43	3.31	14	9	8	13	....
NORTHERN-CENTRAL MD.																
Baltimore		179	39 17	76 36	63.4	72.6	54.3	87 46		41	6.35	9	16	6	15	N. W.
Distributing Res.	Dist. Col	....	38 52	77 0	†64.6	....		96 49		37	4.32	....	....	....	10	....
Fallston	Harford	450	39 31	76 24	*60.2	....		85 41		44	6.10	....	....	....	11	....
Great Falls	Montgomery	....	39 0	77 14	†64.6	....		*6 44		42	3.74	....	....	....	12	....
McDonogh	Baltimore	535	39 23	76 44	62.3	68.4	54.4	80 43		37	3.92	....	....	....	14	....
Mt. St. Mary's	Frederick	720	39 41	77 21	61.8	71.4	52.3	87 41		46	3.17	6	15	10	12	S. W.
New Market	Frederick	500	39 23	77 18	*61.0	....		88 48		40	5.55	11	8	12	12	S. W.
Frederick	Frederick	260	39 24	77 18	63.6	74.1	52.7	87 41		46	3.16	15	6	10	12	....
Receiving Res.	Dist. Col	38	52	77 0	†64.4	....		85 48		37	4.81	....	....	....	10	....
Taceytown	Carrol	....	39 40	77 9	....	....					5.37	....	....	....	14	....
Washington, D. C.		112	38 53	77 0	63.8	73.8	53.9	89 44		45	4.07	12	11	8	14	S.
Woodstock College	Baltimore	392	39 20	76 49	62.1	72.7	51.5	89 39		50	4.78	12	14	5	10	N. W.
SOUTHERN MARYLAND.																
Jewell	Anne Arundel	....	38 44	76 46	†63.7	....		79 51		28	4.75	16	10	5	7	....
Solomon's	Calvert	20	38 19	76 27	64.7	72.8	56.6	84 47		37	2.99	11	8	12	9	S. E.
EASTERN MD. AND DELAWARE.																
Barron Ck. Springs	Wicomico	25	38 30	75 39	63.3	72.6	54.0	85 43		42	3.44	13	13	3	12	S. W.
Easton	Talbot	35	38 42	76 6	64.8	74.5	55.2	89 44		45	5.05	9	18	4	11	S.
Dover, Del	Kent	....	39 9	75 31	61.9	72.2	53.5	86 44		42	5.59	18	7	6	8	S. W
Kirkwood, Del	New Castle	....	39 35	75 40	†62.8	....		82 50		32	....	....	....	....	....	....
Seaford, Del	Sussex	....	38 40	75 35	63.3	73.9	52.7	90 43		47	3.15	....	....	....	11	....
‡ VIRGINIA.																
Norfolk		43	36 51	76 17	67.0	76.6	57.4	98 46		47	3.76	17	11	3	8	S.
AVERAGES	Western Maryland	....	....	....	62.0	71.0	53.1	....		47.0	4.00	14	14	8	14.0	N. W.
	Northern-Centr'l Md.	....	....	....	63.0	72.2	53.2	....		42.3	4.53	10.8	11.7	8.5	12.1	S. W.
	Southern Maryland	....	....	....	64.2	72.8	55.6	....		32.5	3.87	13.5	9.0	7.5	8.0	S. E.
	Eastern Md. and Del.	....	....	....	63.2	74.3	53.8	....		42.0	4.31	14.0	12.7	4.3	14.0	S. W.
	Entire territory	....	....	....	63.1	72.4	53.6	....		41.1	4.37	12.3	11.2	7.4	11.6	S. W.

NOTE.—Letters of the alphabet are used to indicate the number of days that are missing from record: e. g. a—one day, b—two days, etc. (1)—H. SHRIVEL. (2)—E. T. SHRIVEL. * From tri-daily readings. † From bi-daily readings. ‡ Omitted in computing averages.

MONTHLY SUMMARY OF REPORTS FOR JUNE, 1892.

Stations.	Counties.	Altitude above sea in ft.	Latitude.	Longitude.	Monthly Mean.	Mean of Max.	Mean of Min.	Degrees, Max.	Degrees, Min.	Monthly Range.	Total Precipitation.	Clear Days.	Fair Days.	Cloudy Days.	Rainy Days (.01 inch or more.)	Prevailing Wind.
WESTERN MARYLAND.																
Boetteberville......	Allegany ...		29°39′	78°48′	*73.8	81.0	66.6	90	55	35	6.60	....	....	...	13	
Cumberland (1)....	Allegany	650	39 39	78 46	78.7	88.0	69.4	95	59	36	7.31	18	3	9	12	.....
Cumberland (2)....	Allegany	700	39 39	78 45	73.3	80.1	66.6	89	56	33	10.06	....	....		14	....
NORTHERN-CENTRAL MD.																
Baltimore..........		179	39 17	76 36	75.9	84.9	66.9	94	51	40	4.87	8	19	3	13	S. W.
Darlington........	Harford......	300	39 39	76 14	72.8	82.6	63.2	92	51	41	4.02	23	4	3	12	.....
Dist. Res., D. C. ...			38 52	77 00	+72.6	.....	.....	92	58	34	1.42	...	...	...	10	.....
Fallston	Harford	450	39 31	76 24	*72.5	.....	.....	93	54	36	3.35	...	...	...		.....
Frederick........	Frederick....	280	39 24	77 18	76.4	86.0	66.7	94	52	42	2.80	22	3	5	12	.....
Great Falls.......	Montgomery..		39 00	77 14	+77.1	.....	.....	91	60	31	2.10	...	...		8	.....
McDonogh	Baltimore....	535	39 23	76 44	74.8	81.4	68.2	89	53	38	2.90	...	...		10	
Mt. St. Mary's....	Frederick....	720	39 41	77 21	r72.8	80.7	64.6	93	52	41	4.52	...	...		7	.....
New Market....	Frederick.....	500	39 23	77 18	*74.7	.....	.....	98	97	41	3.46	14	14	2	12	S. W.
Rec. Res., D. C.			38 52	77 00	+77.2	.....	.....	90	59	31	3.01	...	...		10	.....
Tanoytown........	Carroll......		39 40	77 9	.....	.....	.....				3.08	...	...		8	.....
Washington, D. C. ..		112	38 52	77 00	76.2	85.2	67.3	94	53	41	2.59	8	19	3	11	S.
Woodstock College.	Baltimore.....	332	39 20	76 49	75.1	84.0	66.2	92	52	40	2.90	9	13	8	6	S. W.
SOUTHERN MARYLAND.																
Jewell	Anne Arundel		38 44	76 40	+76.7	.....	.....	84	65	19	5.68	21	6	3	8	.....
Solomon's.........	Calvert.......	20	39 19	76 27	77.5	83.3	69.6	94	57	37	4.00	6	10	14	8	S. W
EASTERN MD. AND DELAWARE.																
Barron Ck. Springs.	Wicomico.....	25	38 30	75 39	75.6	83.8	67.4	92	51	41	1.82	10	14	6	7	S. W.
Easton..	Talbot	35	38 42	76 6	76.1	84.7	68.6	91	51	40	3.08	12	12	5	5	S. W.
Dover, Del........	Kent..........		39 9	75 31	75.4	83.9	67.0	91	53	38	1.45	21	7	2	7	S. W.
Kirkwood, Del.....	New Castle...		39 35	75 40	+76.1	.....	.....	92	62	33	....	...	...			.....
Seaford, Del.......	Sussex.......		38 40	75 35	71.2	86.2	66.3	95	49	46	2.34	....	....		3	
‡ VIRGINIA.																
Norfolk.		43	36 51	76 17	76.4	84.7	68.0	94	58	36	4.83	14	12	4	14	S.
AVERAGES {	Western Maryland...				75.3	83.0	67.5	....	34.7		8.00	18	3	9	13.0	.....
	Northern-Cent'l Md..				75.2	81.0	66.2	....	37.8		3.15	18.0	12.0	4.0	9.9	S. W.
	Southern Maryland..				77.1	85.3	69.6	....	28.0		4.84	13.5	8.0	8.5	8.0	S. W.
	Eastern Md. and Del.				74.9	84.4	67.3	....	39.6		2.17	14.7	11.0	4.3	5.5	S. W
	Entire territory.....				75.3	83.8	66.9	....	36.9		3.78	14.4	10.3	5.2	9.3	S. W.

NOTE.—Letters of the alphabet are used to indicate the number of days that are missing from record : e. g. a—one day, b—two days, etc. (1)—H. SHRIVER. (2)—E. T. SHRIVER. *From tri-daily readings. +!From bi-daily readings. ‡ Omitted in computing averages.

MONTHLY SUMMARY OF REPORTS FOR JULY, 1892.

STATIONS.	COUNTIES.	Altitude above sea in ft.	Latitude.	Longitude.	Monthly Mean.	Mean of Max.	Mean of Min.	Degrees, Max.	Degrees, Min.	Monthly Range.	Total Precipitation.	Clear Days.	Fair Days.	Cloudy Days.	Rainy Days. (.01 inch or more.)	Prevailing Wind.
WESTERN MARYLAND.																
Boettcherville......	Allegany......		39 39	74 48	*74.6	85.6	63.5	100 50	50	1.10					4	
Cumberland (1)....	Allegany......	650	39 39	78 46	75.6	87.3	64.0	101 52	49	1.22	11	18	2	7	W.
Cumberland (2).....	Allegany......	700	39 39	78 45	72.9	82.7	63.1	97 53	44	1.15	23	3	5	5	
NORTHERN-CENTRAL MD.																
Baltimore...........	... .	179	39 17	76 38	76.4	85.4	67.3	99 58	41	4.07	16	12	3	9	N.W. S.W.
Darlington	Harford.....	300	39 39	76 14	74.2	84.4	64.1	98 55	43	4.53	24	5	2	8	S.W.
Fallston...........	Harford. ...	450	39 31	76 24	*73.2	...	...	94 56	42	5.96	...	...	...	12	....
Frederick..........	Fred rick.....	280	39 24	77 18	76.6	87.2	66.1	99 54	45	2.30	25	4	2	8	....
Great Falls	Montgomery..	...	39 00	77 14	†76.2	...	...	96 58	38	4.19	...	...	...	11	....
McDonogh	Baltimore....	535	39 23	76 44	74.6	83.2	65.9	95 56	39	4.67	...	...	...	6	....
Mt. St. Mary's.....	Frederick.....	730	39 41	77 21	76.6	85.3	67.8	97 54	43	5.31	...	...	...	9	N.W.
New Market........	Frederick.....	500	39 23	77 18	*73.7	...	...	94 56	42	3.89	21	8	2	9	S.W.
Dist. Res., D. C.....	....		38 52	77 00	†76.5	...	...	97 61	36	6.40	...	...	...	11	....
Taneytown........	Carroll ...	...	39 40	77 9	...	...	...	...		4.54	...	...	...	10	....
Washington, D. C..	....	112	38 52	77 00	75.8	85.4	66.1	99 54	45	5.03	16	9	6	10	S.
Rec. Res., D. C.....	....		38 52	77 00	76.2	...	...	94 61	33	4.86	...	...	...	11	....
SOUTHERN MARYLAND.																
Jewell.............	Anne Arundel	...			†76.0	...	...	95 65	29	4.62	22	7	2	6	....
Leonardtown.......	St. Mary's....		38 18	76 40	78.4	86.0	70.7	99 64	35	4.38	...	...	...	9	S.W.
Solomon's.........	Calvert........	20	38 19	76 27	78.2	86.5	69.7	98 61	27	2.49	11	6	14	10	S.W.
EASTERN MD. AND DELAWARE.																
Barren Ck. Springs.	Wicomico. ...	25	38 30	75 39	74.9	83.6	66.2	97 51	46	3.37	8	16	7	11	S.W.
Dover, Del.........	Kent	39	39 9	75 31	75.5	84.6	66.5	100 56	44	4.35	21	6	4	9	S.W.
Easton	Talbot........	35	38 42	76 6	76.6	86.4	66.7	100 56	44	2.93	15	12	4	7	S.W.
Kirkwood, Del......	New Castle...		39 35	75 40	†78.9	...	...	102 60	42	...	...	...	...	...	....
Seaford, Del..	Sussex		38 40	75 35	75.8	85.7	65.8	100 53	47	2.90	...	...	.. ...	8	....
‡ VIRGINIA.																
Norfolk	....	43	36 51	76 17	76.4	84.5	68.2	99 57	42	8.27	10	15	6	14	S.
AVERAGES	Western Maryland...	...	...	...	74.4	85.2	63.6	...	...	47.7	1.16	17.0	10.0	3.5	3.5	W.
	Northern-Cent'l Md..	...	...	...	75.5	85.2	66.2	...	...	40.6	4.72	20.4	7.6	3.0	9.5	S.W.
	Southern Maryland..	...	...	...	77.6	86.2	70.2	.. ..		30.3	3.83	18.5	6.5	9.0	8.5	S.W.
	Eastern Md. and Del.	...	...	...	76.5	85.1	66.3	...	...	45.6	3.31	14.7	11.3	5.0	9.0	S.W.
	Entire territory.....	...	...	...	75.8	85.3	66.2	...	...	41.1	3.81	17.5	8.8	4.4	8.9	S.W.

NOTE.—Letters of the alphabet are used to indicate the number of days that are missing from record: e. g. a—one day, b—two days, etc. (1)—H. SHRIVER. (2)—E. T. SHRIVER. *From tri-daily readings. †From bi-daily readings. ‡Omitted in computing averages.

MONTHLY SUMMARY OF REPORTS FOR AUGUST, 1892.

STATIONS.	COUNTIES.	Altitude above sea in ft.	Latitude.	Longitude.	TEMPERATURE. Monthly Mean.	Mean of Max.	Mean of Min.	Degrees, Max.	Degrees, Min.	Monthly Range.	Total Precipitation.	Clear Days.	Fair Days.	Cloudy Days.	Rainy Days. (.01 inch or more.)	Prevailing Wind.
WESTERN MARYLAND.																
Boettcherville	Allegany.....		39°39'	78°48'	*75.0	84.8	65.3	98	56	42	1.00	....	....	....	4	....
Cumberland (1).....	Allegany.....	650	39 39	78 46	78.2	88.9	67.9	101	59	42	2.03	9	13	9	5	W.
Cumberland (2).....	Allegany.....	700	39 39	78 45	75.2	86.0	64.4	94	58	36	1.90	22	4	5	6	....
Edgemont	Washington ..		39 45	77 29	78.6	90.4	66.7	99	56	43						
NORTHERN-CENTRAL MD.																
Baltimore		179	39 17	76 36	76.2	84.5	67.8	95	60	35	1.93	11	16	4	10	N.W.
Darlington	Harford	300	39 39	76 14	74.0	83.8	64.3	94	57	37	3.39	26	2	3	9	N.W.
Dist. Res., D. C.			38 52	77 0	†77.2			92	62	30	1.40	....	....	....	4	....
Failston	Harford	450	39 31	76 24	*72.1			92	58	34	4.10	....	....	....	8	....
Frederick	Frederick	280	39 24	77 18	76.4	86.4	65.3	95	58	37	1.68	25	6	0	8	....
Great Falls.......	Montgomery .		39 0	77 14	*76.0			94	61	33	.75	....	....	....	3	....
McDonogh	Baltimore ...	535	39 23	76 44	74.1	82.3	65.9	92	59	33	2.39	....	....	....	8	....
Mt. St. Mary's	Frederick	720	39 41	77 21	72.9						3.39	....	....	....	4	N.W.
New Market.......	Frederick ...	500	39 23	77 18	*73.3			94	58	36	2.27	19	11	1	4	W.
Rec. Res., D. C.			38 52	77 0	†76.4			90	64	16	1.19	....	....	....	6	....
Taneytown	Carroll.......		39 40	77 9							.58	....	....	....	2	....
Washington, D. C.		112	38 52	77 0	76.2	85.7	65.8	95	60	35	.85	18	10	3	5	E.
Woodstock College	Baltimore ...	392	39 20	76 49	73.6	82.8	64.3	94	58	36	2.45	14	12	5	3	N.W.
SOUTHERN MARYLAND.																
Jewell	Anne Arundel		38 44	76 36	†77.2			84	69	16	2.47	23	8	0	4	....
Leonardtown	St. Mary's ...		38 18	76 40	76.6	86.1	67.2	94	61	33	1.21	24	3	4	8	S. W.
Solomon's	Calvert	20	38 19	76 27	76.7	85.9	71.5	95	63	32	2.80	12	8	11	8	S. E.
EASTERN MD. AND DELAWARE.																
Barron Ck. Springs.	Wicomico	25	38 30	75 39	76.8	84.4	67.1	92	57	35	2.40	16	9	6	8	S. E.
Dover, Del	Kent		39 9	75 31	75.0	83.0	67.1	92	60	32	2.54	23	6	2	11	S. E.
Easton..........	Talbot	35	38 42	76 6	76.2	86.7	65.6	95	57	38	1.09	18	10	3	3	N. W.
Kirkwood	New Castle		39 35	75 40	†79.7			100	66	34		....	....	....		....
Seaford, Del	Sussex		38 40	75 35	76.0	86.8	66.0	97	59	38	1.39	....	....	....	8	....
‡ VIRGINIA.																
Norfolk		43	36 51	76 17	79.2	87.2	71.2	94	66	28	3.53	18	8	5	10	S. W.
AVERAGES { Western Maryland ..					76.8	87.4	66.1			40.8	1.23	15.5	8.5	7.0	5.0	....
Northern-Cent'l Md.					74.9	84.2	65.9			33.8	2.02	18.7	9.5	2.7	5.7	N.W.
Southern Maryland..					77.5	86.0	69.4			76.7	2.19	19.7	6.3	5.0	5.3	S. E. S. W.
Eastern Md. and Del.					76.5	85.2	66.4			35.4	1.86	19.0	8.3	3.4	10.0	S. E.
Entire territory					75.9	85.5	66.5			34.4	1.96	18.6	8.4	4.7	5.9	....

NOTE.—Letters of the alphabet are used to indicate the number of days that are missing from record: e. g. a—one day, b—two days, etc. (1)—H. SHRIVER. (2)—E. T. SHRIVER. * From tri-daily readings. † From bi-daily readings. ‡ Omitted in computing averages.

MONTHLY SUMMARY OF REPORTS FOR SEPTEMBER, 1892.

STATIONS.	COUNTIES.	Altitude above sea in ft.	Latitude.	Longitude.	TEMPERATURE.						Total Precipitation.	Clear Days.	Fair Days.	Cloudy Days.	Rainy Days. (.01 inch or more)	Prevailing Wind.
					Monthly Mean.	Mean of Max.	Mean of Min.	Degrees, Max.	Degrees, Min.	Monthly Range.						
WESTERN MARYLAND.																
Boetteberville	Allegany		39°39′	78°48′	*64.0	73.7	52.3	88	40	48	2.90					
Cumberland (1)	Allegany	650	39 39	78 46	66.6	78.2	57.0	87	47	40	2.35				5	
Cumberland (2)	Allegany	700	39 39	78 45	64.0	72.8	55.3	83	45	38	2.36	23	2	5	5	W.
Edgemont	Washington		39 45	77 29	69.2	81.8	56.7	96	46	52						
NORTHERN-CENTRAL MD.																
Baltimore		179	39 17	76 36	66.0	74.8	57.7	88	49	39	2.38	19	8	5	9	S. E.
Darlington	Harford	300	39 39	76 14	64.2	74.1	54.3	85	45	40	2.77					E.
Dist. Res., D.C.			39 52	77 0	755.6			84	50	34	4.85				6	
Fall-ton	Harford	450	39 31	76 24	*63.7			85	48	37	3.28				5	
Frederick	Frederick	290	39 24	77 16	65.4	75.4	55.5	88	47	41	5.82	21	5	4	7	
Great Falls	Montgomery		39 0	77 14	*65.3			86	47	39	2.04				7	
McDonogh	Baltimore	555	39 23	76 44	64.4	73.6	55.1	84	49	35	3.32				8	
Mt. St. Mary's	Frederick	720	39 41	77 21	67.2	75.3	59.2	85	48	37	5.52	14	9	7	8	N. W.
New Market	Frederick	500	39 23	77 18	*63.2			86	48	38	3.77	16	5	9	5	W.
Roc. Res., D. C.			38 52	77 0	*66.0			83	51	32	4.10				6	
Taneytown	Carroll		39 40	77 9							2.44				6	
Washington, D. C.		112	38 52	77 0	66.0	76.2	56.0	88	48	40	3.35	15	7	8	8	E.
Woodstock College	Baltimore	392	39 20	76 49	63.5	73.6	53.4	85	44	41	3.53	18	7	5	6	N. W.
SOUTHERN MARYLAND.																
Jewell	Anne Arundel		38 44	76 36	*67.8			80	55	25	3.08	28	1	3	7	
Leonardtown	St. Mary's		38 16	76 40	66.2	75.5	56.8	84	51	33	2.29	22	4	4	4	S. W.
Solomon's	Calvert	20	38 19	76 27	69.2	76.2	62.2	84	54	30	1.75	15	6	9	7	S. E.
EASTERN MD. AND DELAWARE.																
Barron Ck. Springs.	Wicomico	25	38 30	75 39	66.7	76.2	57.2	84	48	36	2.08	18	8	4	3	S. E.
Dover, Del	Kent	39	9	75 31	66.3	75.4	57.3	87	49	38	2.71	25	3	2	4	S. E.
Easton	Talbot	35	38 43	76 6	65.7	75.5	55.9	85	47	38	1.84	17	9	4	8	E.
Kirkwood, Del	New Castle		39 35	75 40	*69.5			88	56	32					5	
Seaford, Del	Sussex		38 40	75 35	68.8	78.2	55.4	90	44	46	1.81				2	
‡ VIRGINIA.																
Norfolk		43	36 51	76 17	71.0	77.0	64.0	84	53	31	1.33	16	9	5	5	N. E.
AVERAGES {	Western Maryland				66.0	76.6	55.3			54.5	2.50	23.0	2.0	5.0	5.0	W.
	Northern-Cent'l Md.				65.0	74.3	55.9			37.8	3.63	17.2	6.5	6.3	7.3	N. W.
	Southern Maryland				64.0	75.8	59.5			29.8	2.37	21.0	3.7	5.3	6.0	S. W. / S. E.
	Eastern Md. and Del.				67.0	76.3	56.4			38.0	2.11	20.0	6.7	3.4	4.6	S. E.
	Entire territory				65.9	75.7	56.3			37.9	3.05	19.1	5.5	5.3	5.8	S. E.

NOTE.—Letters of the alphabet are used to indicate the number of days that are missing from record: *e. g.* a—one day, b—two days, etc. (1)—H. SHRIVER. (2)—E. T. SHRIVER. * From tri-daily readings. † From bi-daily readings. ‡ Omitted in computing averages.

MONTHLY SUMMARY OF REPORTS FOR OCTOBER, 1892.

Stations	Counties	Altitude above sea in ft.	Latitude	Longitude	Monthly Mean	Mean of Max.	Mean of Min.	Degrees, Max.	Degrees, Min.	Monthly Range	Total Precipitation	Clear Days	Fair Days	Cloudy Days	Rainy Days (.01 inch or more.)	Prevailing Wind.
WESTERN MARYLAND.																
Boettcherville......	Allegany.. ...		39°39'	78°48'	*52.9	65.6	40.2	84	30	54	0.20	....	......	...	...	..
Cumberland (1)	Allegany......	650	39 39	78 46	53.4	64.1	42.6	78	34	44	0.27	....	......	...	...	..
Cumberland (2)	Allegany......	700	39 39	78 45	52.6	62.7	42.4	78	30	48	0.24	25	5	1	2	..
NORTHERN-CENTRAL MD.																
Baltimore		179	39 17	76 36	55.8	65.5	48.1	83	34	49	0.26	17	14	0	3	N.W.
Darlington	Harford	300	39 38	76 14	54.3	64.2	44.4	80	31	49	0.38	26	0	5	4	N.W.
Dist. Res., D. C.....			38 52	77 0	*54.8			81	35	46	0.31	....	...	...	2	
Fallston	Harford	450	39 31	76 24	*53.0			80	33	47	0.45	....	......	...	2	
Frederick	Frederick	260	39 24	77 18	54.0	64.5	43.6	82	32	50	0.19	25	6	0	3	
Great Falls	Montgomery		39 0	77 14	*54.1			81	29	52	0.10	....	......	...		
McDonogh........	Baltimore	535	39 23	76 44	53.6	63.2	44.1	78	33	45	0.34	....	...	...	3	
Mt. St. Mary's.....	Frederick	720	39 41	77 21	54.4	65.0	43.7	82	31	51	0.22	17	10	4	3	N.W.
New Market.......	Frederick	500	39 22	77 18	*48.7			75	32	43	0.21	....	...	...		N.W.
Roc. Res., D. C.....			38 52	77 0	*54.4			79	33	36	0.26	....				
Taneytown........	Carroll		39 40	77 9							0.00					
Washington, D. C...		112	38 52	77 0	55.4	66.3	44.4	84	30	54	0.34	17	12	2	3	N.W.
Woodstock College.	Baltimore	392	39 20	76 49	52.0	64.2	39.8	79	26	43	0.24	15	13	3	1	N.W.
SOUTHERN MARYLAND.																
Jewell	Anne Arundel		38 44	76 36	*53.2						0.50	24	7	0	1	
Leonardtown	St. Mary's ..		38 18	76 40	59.2	69.4	49.1	81	34	47	1.13	20	5	6	2	N.W.
Solomon's	Calvert	20	38 19	76 27	58.0	65.0	49.9	84	40	44	0.67	10	16	5	4	N.W.
EASTERN MD. AND DELAWARE.																
Barron Ck. Springs.	Wicomico.....	25	38 30	75 39	54.8	66.7	42.8	81	28	53	0.09	14	14	3	2	N,W.
Dover, Del	Kent.......		39 9	75 31	55.4	65.9	45.0	82	33	49	0.46	24	7	0	5	N.W.
Easton	Talbot	35	38 42	76 6	56.4	68.1	44.7	79	32	47	0.79	17	13	1	3	N.W.
Kirkwood, Del......	New Castle..		39 35	75 40	*54.5			80	36	44		....	......		3	
Seaford, Del........	Sussex		38 40	75 35	55.0	67.0	43.0	84	30	54	0.84	....	...	...		
‡ VIRGINIA.																
Norfolk		43	36 51	76 17	59.1	67.7	50.5	84	37	47	0.52	21	6	4	3	N.
AVERAGES {	Western Maryland ..				53.0	64.1	41.7			48.7	.24					N.W.
	Northern-Cent'l Md.				50.4	66.9	43.7			47.1	.35	19.5	9 2	3.5	2.8	N.W.
	Southern Maryland.				56.8	67.7	49.5			45.5	.77	18.0	9.3	3.7	3.5	N.W.
	Eastern Md. and Del.				55.2	66.9	43.9			49.4	.54	18.3	11.3	1.3	3.2	N.W.
	Entire territory ...				54.3	65.5	44.1			48.1	0.37	19	10	2	3	N.W.

NOTE.—Letters of the alphabet are used to indicate the number of days that are missing from record: e. g. a—one day, b—two days, etc. (1)—H. SHRIVER. (2)—E. T. SHRIVER. *From tri-daily readings. †From bi-daily readings. ‡Omitted in computing averages.

MONTHLY SUMMARY OF REPORTS FOR NOVEMBER, 1892.

STATIONS.	COUNTIES.	Altitude above sea in ft.	Latitude.	Longitude.	TEMPERATURE. Monthly Mean.	Mean of Max.	Mean of Min.	Degrees, Max.	Degrees, Min.	Monthly Range.	Total Precipitation.	Clear Days.	Fair Days.	Cloudy Days.	Rainy Days. (.01 inch or more.)	Prevailing Wind.
WESTERN MARYLAND.																
Boettcherville	Allegany		39°39'	78°46'	*40.0			70	20	50	3.70				11	N.W.
Cumberland (1)	Allegany	650	39 39	78 46	43.0	49.7	36.3	67	24	43	3.58	3	4	23	11	N.W.
Cumberland (2)	Allegany	700	39 39	78 45	41.0	46.0	35.9	65	22	43	3.16	9	11	10	11	
Edgemont	Washington		39 45	77 29	44.5	54.0	35.0	77	15	62						
NORTHERN-CENTRAL MD.																
Baltimore		179	39 17	76 36	43.8	50.2	37.3	70	21	49	3.85	9	14	7	11	N.W.
Darlington	Harford	300	39 39	76 14	41.0	48.8	33.2	69	18	51	4.51	12	8	10	9	N.W.
Denton			38 47	75 41	f 44.1			66	32	34	6.58				8	
Dist. Res., D. C			38 52	77 0	†48.4			58	22	46	4.76					
Fallston	Harford	450	39 31	76 24	*41.5			70	17	53	5.36				12	
Frederick	Frederick	280	39 24	77 18	42.4	44.7	35.2	69	22	47	4.66	19	6	5	11	
Great Falls	Montgomery		39 0	77 14	†43.0			69	22	47	3.11					
McDonogh	Baltimore	535	39 23	76 44	a43.8	52.6	35.0	70	19	51	8.12				8	
Mt. St. Mary's	Frederick	720	39 41	77 21	41.2	48.4	34.0	68	18	50	3.23	6	7	17	10	N.W.
New Market	Frederick	500	39 23	76 27	*41.4			72	20	52	5.36	10	7	13	7	N.W.
Rec. Res., D. C			38 52	77 0	43.2			67	22	45	3.85					
Taneytown	Carroll		39 40	77 9							5.57					
Washington, D. C		112	38 52	77 0	43.7	50.9	30.5	71	22	49	3.38	9	9	12	12	N.W.
Woodstock College	Baltimore	392	39 20	76 49	41.8			69	19	52	4.68	9	10	11	11	N.W.
SOUTHERN MARYLAND.																
Jewell	Anne Arundel		38 44	76 30	†40.8			65	22	43	6.17	12	4	14	11	
Leonardtown	St. Mary's		38 18	76 40	46.8	55.1	38.4	72	21	51	5.31	12	8	10	7	N.W.
Solomon's	Calvert	20	38 19	76 27	46.4	52.9	39.8	72	23	49	3.90	9	6	16	14	N.W.
EASTERN MD. AND DELAWARE.																
Barron Ck. Springs	Wicomico	25	38 30	75 39	a43.4	50.6	36.2	72	24	48	5.39	9	13	8	13	N.W.
Dover, Del	Kent		39 9	75 31	44.6	52.0	37.3	70	22	48	5.79	10	12	8	14	N.W.
Easton	Talbot	35	38 42	76 6	42.8	52.8	33.1	70	22	48	3.10	8	11	11	13	N.W.
Kirkwood, Del	New Castle		39 35	75 40	†41.6			72	22	50					11	
Seaford, Del	Sussex		38 40	75 27	44.6	53.8	35.5	71	22	49	6.60					
‡ VIRGINIA.																
Norfolk		43	36 51	76 17	49.2	56.9	41.4	77	25	50	2.38	12	8	10	14	N.W.
AVERAGES {	Western Maryland				42.1	49.9	35.7			66.0	3.46	6.0	7.5	16.5	11.0	
	Northern-Cent'l Md.				42.6	50.1	35.2			48.0	4.43	10.6	8.7	10.7	10.0	N.W.
	Southern Maryland				44.7	54.0	39.1			47.7	4.79	11.0	5.7	13.3	10.7	N.W.
	Eastern Md. and Del.				43.4	52.5	35.5			48.6	5.22	9.0	12.0	9.0	12.7	N.W.
	Entire territory				42.9	51.2	35.9			48.3	4.31	9.7	8.8	11.7	10.8	N.W.

NOTE.—Letters of the alphabet are used to indicate the number of days that are missing from record: e. g. a—one day, b—two days, etc. (1)—H. SHRIVER. (2)—E. T. SHRIVER. * From tri-daily readings. † From bi-daily readings. ‡ Omitted in computing averages.

MONTHLY SUMMARY OF REPORTS FOR DECEMBER, 1892.

STATIONS.	COUNTIES.	Altitude above sea in ft.	Latitude.	Longitude.	Monthly Mean.	Mean of Max.	Mean of Min.	Degrees, Max.	Degrees, Min.	Monthly Range.	Total Precipitation.	Clear Days.	Fair Days.	Cloudy Days.	Rainy Days. (.01 inch or more.)	Prevailing Wind.
WESTERN MARYLAND.																
Boettcberville..	Allegany		39°39′	78°48′	*31.5			64	6	58	2.10				...	......
Cumberland (1.. ...	Allegany	700	39 39	78 48	c33.6	38.3	27.9	60	12	48	1.64				...	......
Cumberland (2).....	Allegany . ..	700	39 39	78 45	31.6	37.0	26.2	56	9	47	1.73	11	7	13	8	......
NORTHERN-CENTRAL MD.																
Baltimore.............		179	39 17	76 36	33.4	39.0	27.7	64	14	50	2.28	13	8	10	8	N.W.
Darlington..........	Harford......	300	39 39	76 14	30.4	38.3	22.5	61	8	53	2.23		...	...	...	6 N.W.
Dist. Res., D. C.....			38 52	77 0	†32.6			63	10	53	3.02		...		...	......
Fallston..........	Harford......	450	39 31	76 24	*30.7			58	10	48	2.21	9	14	8	6 N.W.
Fenby,......	Carroll	960	39 36	77 5	*31.1			63	8	55	1.60	10	14	7	4 N.W.
Frederick..........	Frederick	280	39 24	77 18	32.4	38.5	26.3	58	11	47	2.04	15	9	7	9 ......
Great Falls........	Montgomery..		39 9	77 14	*32.2			65	7	58	2.94		...		...	......
McDonogh........	Baltimore.....	535	39 23	76 44	c34.4	41.2	27.7	68	14	54	1.97				...	......
Mt. St. Mary's......	Frederick.....	720	39 41	77 21	31.2	38.8	23.8	63	9	54	2.00	13	13	5	8 N.W.
New Market.......	Frederick.....	500	39 23	76 27	*30.4			58	9	49	2.57	15	8	8	7 N.W.
Rec. Res., D. C......			38 52	77 0	†32.4			63	11	52	2.79				...	......
Taneytown........	Carroll		39 40	77 9							1.77				...	......
Washington, D. C....		112	38 52	77 0	33.2	40.2	26.1	67	13	54	2.82	12	7	11	8 N.W.
Woodstock College.	Baltimore.....	392	39 20	76 49	31.0			64	6	58	2.32	12	8	11	11 N.W.
SOUTHERN MARYLAND.																
Jewell.............	AnneArundel		38 44	76 38	†31.4						2.25	15	7	9	6 ......
Leonardtown......	St. Mary's ...		38 18	76 40	34.6			65	14	51	2.50	14	7	10	5 N.W.
Solomon's.	Calvert	20	38 19	76 27	36.0	42.1	29.6	64	16	48	2.44	10	8	13	7 N.
EASTERN MD. AND DELAWARE.																
Barron Ck. Springs.	Wicomico	25	38 30	75 39	33.0	40.7	25.2	67	11	56	2. 3	13	10	8	10 N.W.
Denton.	Caroline	42	38 47	75 41	c33.1			66	12	54	1.93		...		...	......
Dover, Del........	Kent		39 9	75 31	32.9			67	12	55	2.60	15	7	9	9 N.W.
Easton............	Talbot ...	36	38 42	76 6	30.1	34.9	26.3	55	12	43	2.32	11	6	14	6 N.W.
Kirkwood, Del.....	Newcastle ...		39 35	75 40	†29.4			50	6	42	......		...		...	......
Seaford, Del.......	Sussex		38 40	75 27	33.4	42.0	24.8	66	10	56	2.51		...		...	......
‡ NEW JERSEY AND VIRGINIA.																
Penn's Grove, N. J.	Salem.........		39 50	75 33	34.4	40.0	28.9	65	14	51	2.18	9	15	7	7 N.W.
Norfolk, Va		34	36 51	76 17	40.1	46.5	33.8	72	19	53	3.91	11	9	11	8 N.
AVERAGES	Western Maryland..				32.2	37.6	27.0	...		51.0	1.82					
	Northern-Cent'l Md..				36.6	39.3	25.7	...		52.7	2.25	12.5	10.1	8.4	7.5 N.W.
	Southern Maryland..				34.0			...		49.5	2.40	13.0	7.3	10.7	3.0 N. W.
	Eastern Md. and Del..				32.0	39.2	25.1	...		51.0	2.40	13.0	7.7	10.3	8.3 N.W.
	Entire territory......				32.3	39.3	26.3	...		51.8	2.24	12.4	9.3	9.4	7.2 N.W.

NOTE.—Letters of the alphabet are used to indicate the number of days that are missing from record : e. g. a—one day, b—two days, etc. (1)—H. SHRIVER. (2)—E. T. SHRIVER. * From tri-daily readings. † From bi-daily readings. ‡ Omitted in computing averages.

MONTHLY SUMMARY OF REPORTS FOR JANUARY, 1893.

STATIONS.	COUNTIES.	Altitude above sea in ft.	Latitude.	Longitude.	TEMPERATURE. Monthly Mean.	Mean of Max.	Mean of Min.	Degrees, Max. Degrees, Min.	Monthly Range.	Total Precipitation.	Clear Days.	Fair Days.	Cloudy Days.	Rainy Days. (.01 inch or more.)	Prevailing Wind
WESTERN MARYLAND.															
Boettcherville	Allegany	....	39°39'	78°48'	*20.7	....	54 14		68 1.20		....				
Cumberland (1)	Allegany	700	39 39	78 46	b25.3	32.8	17.8 53 3		65 1.30		....				
Cumberland (2)	Allegany	700	39 39	78 46	22.8	28.5	17.0 55 5		60 .72		11	7	13	4	....
Edgemont	Washington	..	39 45	77 29	24.3	32.3	16.3 58 4		62		....				
Sunny Side	Garrett	....	39 22	39 26	*18.1	....	42 12		54 3.50		6	11	14	14	N.W.
NORTHERN-CENTRAL MD.															
Baltimore		179	39 17	76 36	24.6	30.6	18.5 52 1		51 1.78		13	10	8	16	N.W.
Dist. Res., D. C.		..	38 52	77 0	†23.6	....	53 5		58 2.16		....				
Fallston	Harford	450	39 31	76 24	†22.0	....	46 3		49 2.43		11	13	7	5	N.W.
Fenby	Carroll	950	39 33	77 5	†20.5	....	51 8		59 2.31		8	15	3	8	N.W.
Frederick	Frederick	290	39 24	77 18	22.7	28.5	16.9 47 3		50 1.82		17	7	7	8	....
Glyndon	Baltimore	660	39 27	76 41	19.6	....	....		1.19		11	11	9	6	N.W.
Great Falls	Montgomery	..	39 9	77 14	†23.8	....	53 8		61 1.84		....				
McDonogh	Baltimore	535	39 23	76 44	22.6	30.3	14.8 49 3		52 .84		....				
Mt. St. Mary's	Frederick	730	39 41	77 21	21.2	28.2	14.1 48 6		54 2.36		5	10	16	7	N.W.
New Market	Frederick	50	39 23	78 27	*20.9	....	49 4		53 2.40		14	7	10	5	N.W.
Rec. Res., D. C.		..	38 52	77 0	†24.0	....	52 8		60 2.19		....				
Taneytown	Calvert	..	39 40	77 9	....	....	....		1.96		....				
Washington, D. C.		112	38 52	77 0	24.6	32.5	16.8 57 6		63 1.85		12	10	9	10	N.W.
Westminster	Carroll	..	39 35	77 3	*19.8	....	46 7		54 1.73		10	12	9	8	N.W.
Woodstock College	Baltimore	392	39 20	76 49	22.8	....	49 14		63 2.07		4	17	10	7	N.W.
SOUTHERN MARYLAND.															
Jewell	Anne Arundel	....	38 44	76 36	24.6	....			2.45		19	0	12	7	
Leonardtown	St. Mary's	....	38 18	76 40	25.6	34.4	16.9 58 2		60 2.18		17	7	7	6	N.W.
Solomon's	Calvert	20	38 19	76 27	26.2	33.8	18.5 50 4		48 1.53		4	15	12	7	N.W.
EASTERN MD. AND DELAWARE.															
Barron Ck. Springs	Wicomico	25	38 30	75 38	23.5	32.8	14.2 56 10		66 1.64		9	17	5	13	N.W.
Cambridge	Dorchester	..	38 33	76 6	27.0	33.7	20.4 53 4		35 2.71		19	0	12	7	N.W.
Deoton	Harford	42	38 47	75 41	d24.5	....	56 17		73 1.90		....				
Dover, Del	Kent	..	39 9	75 31	23.9	32.2	15.6 54 5		59 2.39		17	10	4	12	N.W.
Easton	Talbot	35	38 42	76 6	23.9	30.6	17.2 52 1		53 1.29		18	7	6	7	N.W.
Kirkwood, Del	New Castle	..	39 35	75 40	†20.8	....	50 4		54		....				
Millsboro, Del	Sussex	..	38 35	75 15	23.2	34.1	12.4 54 17		71 2.18		18	6	7	8	N.
Salisbury	Wicomico	..	38 26	75 35	†22.7	....	45 7		52 2.21		17	0	14	13	N.W.
Seaford, Del	Sussex	..	38 40	75 35	24.4	34.8	14.1 57 5		62 2.13		....				
‡ NEW JERSEY AND VIRGINIA.															
Penn's Grove, N. J.	Salem	..	39 50	75 33	25.0	33.1	18.0 52 3		55 5.23		8	16	7	8	N. E.
Norfolk, Va		34	36 51	76 17	30.5	38.7	22.3 67 6		2.55		12	9	10	11	N.W.
AVERAGES {	Western Maryland	....	....	....	21.2	31.2	17.0		61.0 1.68		8.5	9.0	13.5	9.0	N.W.
	Northern-Cent'l Md	....	....	....	22.5	30.0	18.2		55.8 1.93		10.5	11.2	8.8	8.0	N.W.
	Southern Maryland	....	....	....	25.5	34.1	17.7		53.0 1.92		13.3	7.9	10.3	6.7	N.W.
	Eastern Md. and Del	....	....	....	23.8	33.0	15.6		57.2 1.83		16.3	6.7	6.5	9.5	N.W.
	Entire territory	....	....	....	22.2	31.9	16.4		56.0 1.94		12.1	9.5	9.1	8.4	N.W.

NOTE.—Letters of the alphabet are used to indicate the number of days that are missing from record: e. g. a—one day, b—two days, etc. (1)—H. SHRIVER. (2)—E. T. SHRIVER. * From tri-daily readings. † From bi-daily readings. ‡ Omitted in computing averages.

MONTHLY SUMMARY OF REPORTS FOR FEBRUARY, 1893.

STATIONS.	COUNTIES.	Altitude above sea in ft.	Latitude.	Longitude.	TEMPERATURE. Monthly Mean.	Mean of Max.	Mean of Min.	Degrees, Max.	Degrees, Min.	Monthly Range.	Total Precipitation.	Clear Days.	Fair Days.	Cloudy Days.	Rainy Days. (.01 inch or more.)	Prevailing Wind.
WESTERN MARYLAND.																
Boettcherville	Allegany		39°39'	78°48'	*31.6			58	5	63	4.40					N.W.
Cumberland (1)	Allegany	700	39.39	78 46	32.8	40.2	25.3	57	2	55	3.96					
Cumberland (2)	Allegany	700	39.39	78 46	31.8	38.1	25.5	58	3	55	3.58	8	11	9	8	
Edgemont	Washington		39 45	77 29	31.5	39.0	24.0	59	4	55						
Sunny Side	Garrett		39 22	39 26	*24.4						4.34	5	13	10	16	S. W.
NORTHERN-CENTRAL MD.																
Baltimore		178	39 17	76 36	34.0	40.9	27.2	61	11	50	4.43	9	10	9	14	N.W.
Darlington	Harford	300	39 39	76 14	31.0	39.6	22.5	58	2	56	4.74	10	8	8	9	N.W.
Dist. Res., D. C.			38 52	77 0	*35.0			81	13	48	3.53					
Fallston	Harford		39 31	76 24	*31.3			59	6	53	5.43	6	13	9	9	N.W.
Fenby	Carroll	950	39 33	77 5	*29.6			56	4	52	4.24	5	12	11	7	N.W.
Frederick	Frederick	280	39 24	77 18	32.6	39.2	26.1	58	10	48	4.06	12	11	5	12	
Glyndon	Baltimore	660	39 27	76 41	30.8			57	7	50	4.51	8	11	9	15	N.W.
Great Falls	Montgomery		39 9	77 14	†34.4			60	9	51	2.76					
McDonogh	Baltimore	535	39 23	76 44	33.2	41.9	24.5	58	5	53	4.14					
Mt. St. Mary's	Frederick	720	39 41	77 21	31.1	38.9	23.3	61	5	56	4.70	5	13	10	10	N.W.
New Market	Frederick	500	39 23	76 27	*31.2			58	6	52	5.01	7	11	10	12	N.W.
Rec. Res., D. C.			38 52	77 0	34.5			60	12	48	3.67					
Woodstock College	Baltimore	392	39 20	76 49	33.2			58	4	54	5.48	5	12	11	18	N.W.
Westminster	Carroll		39 35	77 3	*30.1			57	6	51	4.30	6	13	9	8	N.W.
Taneytown	Calvert		39 40	77 9							2.60					
Washington, D. C.		112	38 52	77 0	35.2	42.9	27.4	65	11	35	4.25	9	7	12	13	N.W
SOUTHERN MARYLAND.																
Jewell	Anne Arundel		35 44	76 36	†35.2						2.22	8	9	11	6	N.W.
Solomon's	Calvert	20	38 19	76 27	36.4	43.8	29.1	65	16	49	4.65	4	3	21	14	N.W.
Leonardtown	St. Mary's		38 18	76 40	36.3	43.7	29.9	66	15	51	4.60	9	11	8	8	N. E.
EASTERN MD. AND DELAWARE.																
Barron Ck. Springs.	Wicomico	25	38 30	75 39	35.2	42.8	27.6	63	12	51	4.39	6	13	9	11	N.W.
Cambridge	Dorchester		39 33	76 6	38.9	47.9	30.0	69	16	53	4.80	12	1	15	7	
Dover, Del	Kent		39 9	75 31	34.9	43.0	26.8	64	11	53	5.05	9	8	14	16	N.W.
Easton	Talbot	35	38 42	76 6	34.8	42.4	28.8	63	11	52	3.87	9	11	8	7	N.W.
Kirkwood, Del	New Castle		39 35	75 40	†28.2			58	8	50						
Denton	Caroline	42	38 47	75 41	†36.4			70	12	58	4.38					
Millsboro, Del	Sussex		38 35	75 15	36.8	46.8	26.9	68	12	54	4.43	13	7	8	11	N.
Salisbury	Wicomico		38 26	75 35	*36.4			60	17	43		17	1	10		N.W.
Seaford, Del	Sussex		38 40	75 35	30.8	47.2	28.4	68	13	53	6.85					
‡ VIRGINIA.																
Norfolk			34 36	51 76 17	41.9	49.6	34.2	70	23	47	4.04	10	7	11	11	N. E.
	Western Maryland				30.4	39.1	24.9			51.0	4.22	6.5	12.0	9.5	12.0	N.W. S.W.
AVERAGES	Northern-Cent'l Md.				32.5	40.6	25.2			51.1	4.18	7.5	11.0	9.4	11.6	N.W.
	Southern Maryland				36.5	43.8	29.5			50.0	3.82	7.0	7.7	13.3	9.3	N.W.
	Eastern Md. and Del.				35.9	45.1	27.8			51.9	4.6	11.0	6.8	10.7	10.2	N.W.
	Entire territory				33.3	42.3	26.4			52.4	4.23	8.3	9.8	10.3	11.0	N.W.

NOTE.—Letters of the alphabet are used to indicate the number of days that are missing from record; e. g. a—one day, b—two days, etc. (1)—H. SHRIVER. (2)—E. T. SHRIVER. *From tri-daily readings. †From bi-daily readings. ‡ Omitted in computing averages,

MONTHLY SUMMARY OF REPORTS FOR MARCH, 1893.

STATIONS.	COUNTIES.	Altitude above sea in ft.	Latitude.	Longitude.	TEMPERATURE. Monthly Mean.	Mean of Max.	Mean of Min.	Degrees Max., lowest Min.	Monthly Range.	Total Precipitation	Clear Days.	Fair Days.	Cloudy Days.	Rainy Days. (.01 inch or more.)	Prevailing Wind.
WESTERN MARYLAND.															
Boettcherville	Allegany		39 39	78 48	*39.3			64 16	48 2.40						
Cumberland	Allegany	650 39 39	78 46	42.6	a51.1	34.1 69 18	50 1.26								
‡ Edgemont	Washington	1100 39 43	77 29	37.3	43.6	30.8 80 10	50								N. W.
Cumberland	Allegany	700 39 39	78 45	39.2	46.9	31.6 64 16	48 1.00	12	9	10	3				
Sunny Side	Garrett	39 22	79 26	*34.0			70 6	64 1.24	11	11	9	10	W.			
NORTHERN-CENTRAL MD.															
Baltimore		178 39 17	76 36	40.3	47.6	33.0 62 16	46 1.38	11	11	9	11	N. W.			
Darlington	Harford	300 39 39	76 14	38.0	47.0	29.0 62 12	50 2.10	14	5	12	7	W.			
Dist. Res., D. C.		38 52	77 0	†40.6			61 17	44 3.32								
Fallston	Harford	450 39 31	76 24	*37.7			60 11	49 1.35	7	21	3	9	N. W.			
Fenby	Carroll	850 39 33	77 5	*35.5			61 12	49 1.65	8	13	10	6	N. W.			
Frederick	Frederick	240 39 54	77 18	39.6	47.0	32.1 60 16	44 1.64	11	16	4	8				
Glyndon	Baltimore	39 39 0	77 14	37.4	46.0	28.8 60 10	50 1.67	15	6	10	10	N. E.			
Great Falls	Montgomery	39 0	77 14	†40.0			60 18	42 1.70								
Mt. St. Mary's	Frederick	730 39 41	77 21	38.2	47.2	29.2 6 12	50 2.42	9	14	8	7	N. W.			
New Market	Frederick	540 39 23	77 18	*36.9			62 16	46 1.90	5	7	19	5	N. W.			
Rec. Res., D. C.		38 52	77 0	40.4			60 17	43 2.26								
McDonogh	Baltimore	535 39 23	76 44	40.0	47.5	32.4 62 15	47 2.24								
Taneytown	Carroll	39 40	77 9					1.99								
Washington, D. C.		112 38 52	77 0	41.0	49.7	32.4 64 17	47 1.83	11	8	12	12	N. W.			
Westminster	Carroll	39 35	77 3	*36.4			58 13	45 1.35	12	10	9	4	N. W.			
Woodstock College	Baltimore	392 39 20	76 49	39.2	48.3	30.1 63 13	50 1.52	10	11	10	8	N. W.			
SOUTHERN MARYLAND.															
Jewell	Anne Arundel	38 44	76 36	42.4				2.80								
Leonardtown	St. Mary's	38 18	76 40	42.5	51.9	33.1 72 16	56 2.68								
Solomon's	Calvert	20 38 19	76 27	42.2	50.5	33.9 70 18	52 2.87	8	6	17	8	S. E.			
EASTERN MD. AND DELAWARE.															
Barron Ck. Springs	Wicomico	25 38 30	75 39	40.4	49.1	31.8 73 15	58 3.85	14	9	8	19	N. W.			
Cambridge	Dorchester	39 33	76 46	44.4	52.2	36.6 68 19	49 4.14	15	5	11	8	W.			
Denton	Caroline	42 38 47	75 41	a43.2	54.1	32.2 70 15	55 3.27	19	3	9	5	W.			
Dover, Del.	Kent	39 8	75 31	40.2	49.2	31.2 71 13	58 3.34	16	7	8	10	N. W.			
Easton	Talbot	35 38 42	76 6	41.6	51.1	32.0 66 15	51 3.44	12	14	5	8	N. W.			
Kirkwood, Del.	New Castle	39 35	75 40	*35.8												
Millsboro, Del.	Sussex	38 35	75 15	40.0	50.1	30.2 72 12	60 3.09	12	5	14	12	N.			
Salisbury	Wicomico	38 26	75 35	*39.0												
Seaford, Del.	Sussex	38 40	75 35	41.2	52.0	30.3 72 12	60 3.28								
‡ VIRGINIA.															
Norfolk		43 36 51	76 17	46.0	53.0	38.0 78 24	54 3.32	13	5	13	13	N. E.			
AVERAGES {	Western Maryland				38.5	47.3	32.2	52.0 1.48	10.2	10.9	9.5	6.5	W. N. W.			
	Northern-Cent'l Md.				39.0	48.3	31.0	47.3 2.00	11.1	11.2	9.6	7.5	N. W.			
	Southern Maryland				42.4	51.0	38.5	54.0 2.78	8.0	6.0	17.0	8.0	S. E.			
	Eastern Md. and Del.				40.8	50.6	32.0	56.0 3.44	13.8	8.0	9.2	12.6	N. W.			
	Entire territory				39.7	49.4	31.8	50.4 2.30	11.6	9.6	9.8	8.7	N. W.			

NOTE.—Letters of the alphabet are used to indicate the number of days that are missing from record: a, g, a—one day, b—two days, etc. (1)—H. SHRIVER. (2—E. T. SHRIVER. * From tri-daily readings. † From bi-daily readings. ‡ Omitted in computing averages.

MONTHLY SUMMARY OF REPORTS FOR APRIL, 1893.

STATIONS.	COUNTIES.	Altitude above sea in ft.	Latitude.	Longitude.	Monthly Mean.	Mean of Max.	Mean of Min.	Degrees, Max.	Degrees, Min.	Monthly Range.	Total Precipitation.	Clear Days.	Fair Days.	Cloudy Days.	Rainy Days. (.01 inch or more)	Prevailing Wind.
WESTERN MARYLAND.																
Boettcherville	Allegany		39°33'	78°48'	†51.0			92	34	58	4.10					
Cumberland (1)	Allegany	650	39 39	78 48	55.3	82.7	48.0	86	87	48	3.90	8	0	22	9	
Cumberland (2)	Allegany	650	39 39	78 45	53.3	80.1	44.6	88	32	56	3.90	15	6	10	14	
Edgemont	Washington	35	39 31	77 38	50.6	69.0	42.3	76	31	45						
Sunny Side	Garrett	20	39 40	79 27	*46.2			81	22	59		7	2	21	18	S. W.
Oakland	Garrett	500	39 52	77 29	47.4			81	28	53	6.75	10	8	12	19	S. W.
Miliville, N. J	Cumberland				54.0	63.4	44.7	76	34	42	3.40	9	11	10	10	N. W.
NORTHERN-CENTRAL MD.																
Baltimore		179	39 17	76 38	52.7	60.6	44.6	81	36	45	3.52	9	11	10	15	S. E.
Darlington	Harford	300	39 47	76 14	49.9	59.1	40.7	78	32	46	3.18	16	5	9	9	S. W.
Dist. Res., D. C.			39 9	77 0	†54.0			75	34	37	2.85					N.W.
Fallston	Harford	1100	39 33	76 24	*50.5			78	35	43	4.00	0	15	15	13	E.
Fenby	Carroll	450	39 24	77 5	*50.6			78	35	43	4.90	7	11	12	11	S. W.
Frederick	Frederick	250	39 27	77 18	52.7	63.0	43.9	77	34	44	4.89	9	14	7	12	
Glyndon	Baltimore	280	39 0	77 41	50.0	58.7	41.4	78	33	45	4.63	15	5	10	14	S. E.
Great Falls	Montgomery	760	38 44	77 14	†54.0			77	30	38	2.69					
Mt. St. Mary's	Frederick		39 23	77 21	51.6	61.1	42.1	79	33	46	3.61	5	13	12	10	S. E.
New Market	Frederick	720	39 25	77 18	*51.2			78	36	42	3.36	7	13	11	8	N.W.
McDonogh	Baltimore		38 35	76 44	51.0	58.6	43.4	76	34	42	3.72					
Rec. Res., D. C.		2375	38 40	77 0	†54.0			76	39	37	2.63					
Taneytown	Carroll		38 52	77 9							5.38					
Washington, D. C.			39 20	77 0	53.8	63.2	44.4	84	35	49	3.21	7	13	10	15	S.
Woodstock	Howard	112	36 51	76 49	51.6	61.1	42.1	79	33	45	3.61	5	13	12	10	S. B.
SOUTHERN MARYLAND.																
Jewell	Anne Arundel		39 35	76 36	†56.0						2.56	10	12	8	11	N. E.
Leonardtown	St. Mary's		39 23	76 40	52.6	61.6	43.5	82	36	46	4.16	10	13	7	11	S. E.
Solomon's	Calvert		39 22	76 27	54.8	63.5	46.0	81	32	49	3.54	4	5	21	12	S. E.
Benedict	Charles		38 31	76 39	58.6	66.1	51.1	81	42	39	3.28	18	0	12	5	W.
EASTERN MD. AND DELAWARE.																
Barron Ck. Springs.	Wicomico	25	38 30	75 39	53.1	59.1	46.6	75	35	40	5.42					
Cambridge	Dorchester		39 39	76 46	57.6	65.6	49.6	78	40	38	3.47	11	6	14	9	
Dover, Del	Kent		38 42	75 31	53.4	62.9	43.9	77	37	49	3.59	12	8	10	16	S. W.
Easton	Talbot	35	49 45	76 6	55.0	65.4	44.7	80	37	43	3.38	6	16	8	9	N.W.
Denton	Caroline	72	38 52	75 41	50.0	58.6	41.1	82	34	49	3.77	8	13	9	8	
Kirkwood, Del	New Castle		38 18	75 40	†48.5											
Millsboro, Del	Sussex	535	39 41	75 15	53.6	64.0	43.3	80	34	46	5.31	11	9	10	13	N.
Seaford, Del	Sussex		38 19	75 35	54.4	66.0	47.8	78	35	43	4.68					
‡ VIRGINIA.																
Cape Charles	Northampton				56.2			81	40	41	2.42	4	11	15	11	S.
Birdsnest	Northampton				57.3			84	42	42	2.10					
Norfolk					60.0	69.0	50.0	87	40	47	2.07	11	14	5	11	S. W.
Warsaw	Richmond				59.3	72.8	45.8	89	34	55	3.39	4	19	7	11	E.
AVERAGES	Western Maryland				51.0	61.3	44.9			61.7	6.22	9.8	6.2	21.0	13.6	S. W.
	Northern-Cent'l Md.				52.0	60.6	42.8			43.1	3.76	8.0	11.2	11.1	11.7	S. E.
	Southern Maryland				55.5	63.7	46.9			44.7	3.38	10.5	7.5	9.5	9.3	S. E.
	Eastern Md. and Del.				53.3	63.1	51.7			42.7	4.23	9.6	10.2	10.2	11.0	
	Entire territory				52.4	61.9	44.3			45.2	3.94	9.4	9.0	11.8	11.7	S. E.

NOTE.—Letters of the alphabet are used to indicate the number of days that are missing from record : e. g. a—one day, b—two days, etc. (1)—H. SHRIVER. (2)—E. T. SHRIVER. * From tri-daily readings † From bi-daily readings. ‡ Omitted in computing averages.

MONTHLY SUMMARY OF REPORTS FOR MAY, 1893.

STATIONS.	COUNTIES.	Altitude above sea in ft.	Latitude.	Longitude.	TEMPERATURE.					Total Precipitation.	Clear Days.	Fair Days.	Cloudy Days.	Rainy Days. (.01 inch or more.)	Prevailing Wind.
					Monthly Mean.	Mean of Max.	Mean of Min.	Degrees, Max. Degrees, Min.	Monthly Range.						
WESTERN MARYLAND.															
Boettcherville	Allegany	....	39 33	78 48	*61.5	....	..	92 40	52	4.90	..	..	..	..	......
Cumberland (1)	Allegany	650	39 39	78 46	64.3	74.1	54.3	90 44	46	4.50	..	..	..	..	......
Cumberland (2)	Allegany	650	39 39	78 45	61.4	70.1	52.8	88 42	46	4.37	12	6	13	8	......
Oakland	Garrett	2375	39 24	79 29	*55.1	....	..	79 38	41	4.78	10	16	5	16	S. W.
Sunny Side	Garrett	....	39 20	79 28	*53.2	....	..	79 33	46	6.58	..	..	..	..	S. W.
NORTHERN-CENTRAL MD.															
Baltimore		179	39 17	76 35	61.4	70.1	52.7	89 45	44	3.78	14	8	9	14	W.
Darlington	Harford	300	38 47	76 14	59.0	69.4	48.6	86 40	46	3.97	14	11	6	6	....
Dist. Res., D. C.		..	38 9	77 0	†62.3	....	..	86 45	41	4.81	..	..	..	..	......
Fallston	Harford	450	39 31	76 24	*59.1	....	..	88 44	44	5.66	..	..	..	..	......
Fenby	Carroll	960	39 35	77 5	*59.2	....	..	82 44	38	6.00	10	13	8	8	N. W.
†Frederick	Frederick	290	39 24	77 18	61.4	72.1	50.8	89 40	49	4.78	13	15	3	12	....
Glyndon	Baltimore	280	39 24	77 41	59.8	69.2	50.4	88 41	47	5.08	18	8	5	12	S. W.
Great Falls	Montgomery	..	39 0	77 14	†62.6	....	..	87 45	42	3.44	..	..	..	..	......
McDonogh	Baltimore	538	39 23	76 44	59.2	67.8	50.6	85 42	43	4.07	..	..	..	..	......
Mt. St. Mary's	Frederick	720	39 41	77 21	60.2	69.9	50.5	87 40	47	4.24	13	6	12	12	N.
New Market	Frederick	500	39 23	77 18	*62.0	....	..	88 44	44	5.20	16	7	8	8	N. W.
Rec. Res., D. C.		..	38 52	77 0	†62.6	....	..	85 46	49	4.77	..	..	..	..	......
Taneytown	Carroll	..	38 40	77 9	....	....	..	....	..	3.66	..	..	..	..	......
Washington, D. C.		112	38 52	77 0	61.6	71.9	51.2	89 40	49	5.41	11	13	7	14	S.
Woodstock College.	Baltimore	392	39 20	76 49	59.6	70.8	48.5	88 38	50	5.15	11	8	12	7	N. W.
SOUTHERN MARYLAND.															
Benedict	Charles	..	38 31	76 39	63.9	73.3	53.3	91 45	48	4.14	20	0	11	7	......
Solomon's	Calvert	20	38 19	76 27	63.4	72.6	54.2	87 47	40	4.01	11	3	17	12	S. E.
Upper Marlboro	Pr. George's	..	38 47	76 45	61.8	74.1	49.6	90 38	52	4.44	17	0	14	9	S. E.
Jewell	Anne Arundel	..	38 44	76 36	†63.8	....	..	....	..	4.52	..	..	..	..	S. E.
Leonardtown	St. Mary's	..	38 18	76 40	62.8	71.0	52.5	84 35	49	4.71	21	7	3	7	S. E.
EASTERN MD. AND DELAWARE.															
Barron Ck. Springs.	Wicomico	35	38 30	75 39	61.3	71.4	51.2	88 40	46	4.29	..	..	..	..	S. E.
Cambridge	Dorchester	..	39 39	76 46	66.2	73.8	58.7	88 51	37	4.41	..	..	..	..	......
‡Denton	Caroline	72	38 52	75 41	†61.2	72.2	50.2	88 37	51	3.49	..	..	..	..	......
Dover, Del	Kent	..	38 42	75 31	61.4	70.8	52.0	90 42	48	3.50	17	9	5	9	S. E.
Easton	Talbot	35	38 42	76 6	63.6	74.6	52.7	87 43	44	4.16	16	9	6	9	W.
Kirkwood, Del	New Castle	..	39 35	75 40	†58.0	....	..	....	..	....	..	..	..	..	......
Milford, Del	Kent	..	38 46	75 25	62.8	73.6	51.9	89 41	48	3.08	23	6	2	10	W.
Millsboro, Del	Sussex	..	39 41	75 15	61.2	72.3	50.3	92 37	55	3.54	17	7	7	11	N.
Seaford, Del	Sussex	..	38 40	75 35	62.4	74.5	50.4	89 36	53	3.80	..	..	..	..	......
‡ VIRGINIA.															
Birdnest	Northampton	..	....	.....	65.4	....	..	90 52	38	4.40	13	8	10	7	N. E.
Cape Charles	Northampton	..	....	.....	64.4	....	..	....	..	3.87	..	..	..	..	......
Norfolk		..	....	.....	66.0	75.0	57.0	90 49	41	6.79	15	10	6	11	S. W.
Warsaw	Richmond	..	....	.....	63.5	76.3	50.7	90 40	50	4.37	8	17	6	8	S.
AVERAGES { Western Maryland		..	..	.....	59.1	72.1	53.5	....	..	46.2	5.03	11.0	11.0	9.0	12.0 S. W
Northern-Cent'l Md.		..	..	.....	60.7	70.1	50.4	....	..	45.9	4.74	13.3	9.9	7.8	10.3 N. W.
Southern Maryland		..	..	.....	63.0	72.8	52.4	....	..	47.2	4.36	17.2	2.5	11.0	8.8 S. E.
Eastern Md. and Del		..	..	.....	62.3	73.9	52.1	....	..	49.0	3.69	18.2	7.8	5.0	10.5 S. E.
Entire territory		..	..	.....	61.2	71.8	51.7	....	..	46.2	4.78	15.1	8.0	8.3	10.1 S. E.

NOTE.—Letters of the alphabet are used to indicate the number of days that are missing from record: e. g a—one day, b—two days, etc. (1)—H. SHRIVER. (2)—E. T. SHRIVER. *From tri-daily readings. †From bi-daily readings. ‡ Omitted in computing averages.

MONTHLY SUMMARY OF REPORTS FOR JUNE, 1893.

STATIONS.	COUNTIES.	Altitude above sea in ft.	Latitude.	Longitude.	Monthly Mean.	Mean of Max.	Mean of Min.	Degrees, Max.	Degrees, Min.	Monthly Range.	Total Precipitation.	Clear Days.	Fair Days.	Cloudy Days.	Rainy Days. (.01 inch or more.)	Prevailing Wind.
WESTERN MARYLAND.																
Boettcherville	Allegany	....	39°33'	78°48'	*78.4	81.6	65.2	98	58	40	3.40	...	...	...	9	..
Cumberland (1).	Allegany	650	39 39	78 46	b75.1	b85.0	a65.1	b96	67	3	1.86	...	...	..	6	E.
Cumberland (2).	Allegany	850	39 39	78 45	70.9	79.4	62.4	91	55	36	2.12	14	10	6	6	..
Oakland	Garrett	2376	39 24	79 29	*67.7	76.0	....	86	54	32	2.73	14	15	1	12	S. W.
Sunny Side.	Garrett	...	39 20	79 28	65.0	77.7	52.4	91	40	51	3.95	11	11	8	12	E.
NORTHERN-CENTRAL MD.																
Baltimore		179	39 17	76 38	72.4	80.8	64.1	96	57	41	2.26	9	11	10	14	E.
Darlington	Harford	300	39 47	76 14	70.4	79.9	60.8	94	50	44	4.74	16	5	9	10	N. E.
Dist. Res., D. C.		...	39 9	77 0	*73.2	....	....	92	60	32	1.86	...	...	...	7	..
Fallston	Harford	450	39 31	76 24	*69.8	....	....	94	58	38	4.51	...	...	...	8	E.
Fonby	Carroll	950	39 33	77 5	*69.6	....	....	93	54	39	5.00	...	...	...	7	N. W.
Frederick	Frederick	280	39 24	77 18	73.0	83.0	63.1	96	55	41	1.48	16	11	8	10	..
Glyndon	Baltimore	280	39 24	77 41	69.8	78.0	61.3	96	52	44	2.39	15	5	10	10	S. E.
Great Falls	Montgomery	...	39 0	77 14	†72.9	....	....	93	58	35	1.65	...	...	...	9	..
McDonogh	Baltimore	585	39 23	76 44	b70.2	b78.3	b62.2	b94	b55	38	3.13	...	...	...	10	..
Mt. St. Mary's.	Frederick	720	39 41	77 21	72.3	80.8	63.6	95	50	45	2.48	9	6	15	14	N.
New Market.	Frederick	500	39 32	77 18	*71.6	....	....	95	58	37	3.50	12	12	6	8	N. W.
Rec. Res., D. C.		...	38 52	77 0	†74.2	....	....	91	63	31	1.68	...	...	...	6	..
Washington, D. C.		112	38 52	77 0	72.5	81.5	63.5	96	51	41	1.81	11	11	8	11	N. W.
Woodstock College	Baltimore	392	39 20	76 49	a71.0	80.4	a61.7	95	a53	42	3.39	10	11	9	8	N. R.
SOUTHERN MARYLAND.																
Benedict	Charles	...	38 31	76 39	75.1	85.2	65.0	90	55	44	1.30	16	12	0	3	E.
Charlotte Hall	St. Mary's	...	38 30	76 44	L76.5	L68.8	L64.2	L100	L51	49	1.08	...	...	...	9	S. E.
Jewell	Anne Arundel	...	38 44	76 36	†74.3	....	....	....	....	..	1.04	19	5	6	4	S. E.
Leonardtown	St. Mary's	...	38 18	76 40	72.8	81.8	64.3	94	54	40	2.09	14	10	6	4	S. W.
Solomon's	Calvert	20	38 19	76 27	74.0	82.4	65.7	95	59	36	2.67	9	7	14	9	S. W.
Upper Marlboro	Pr. George's.	...	38 47	76 45	73.0	83.2	62.8	97	52	45	1.67	10	11	9	6	N. E.
EASTERN MD. AND DELAWARE.																
Barren Ck. Springs	Wicomico	25	38 29	75 39	71.5	80.7	62.3	93	48	45	.72	12	18	5	6	N. E. N. S.
Cambridge	Dorchester	...	39 39	76 46	76.6	83.1	70.0	96	61	35	2.94	1	18	11	6	S. N. W.
Denton	Caroline	72	38 52	75 41	72.4	80.5	64.2	94	54	40	1.80	0	23	7	4	W.
Dover, Del	Kent	...	38 42	75 31	71.5	80.2	62.8	96	51	45	1.20	14	11	5	6	E.
Easton	Talbot	35	38 42	76 5	73.0	82.5	63.4	95	53	42	1.46	13	11	6	5	N. E.
Kirkwood, Del.	New Castle	...	39 35	75 40	†72.0	....	....	104	58	..	....	...	...	...	8	..
Milford, Del	Kent	...	38 56	75 25	72.4	81.1	63.7	95	55	40	1.69	22	..	7	6	N. E.
Millsboro, Del	Sussex	...	39 41	75 15	71.6	81.7	61.4	97	47	50	2.23	14	12	4	6	S.
Seaford, Del	Sussex	...	38 40	75 35	71.8	82.8	60.7	97	52	45	.77	...	...	...	4	..
‡ VIRGINIA.																
Birdsnest	Northampton	...	...	...	*76.2	....	....	92	62	30	5.96	9	12	9	5	N. E. S. W.
Cape Charles	Northampton	...	...	...	*72.3	....	61.8	89	49	40	6.78	c15	c0	c12	4	S. W.
Norfolk		...	...	...	78.9	81.7	66.1	91	59	34	8.36	10	16	4	9	N. E.
Warsaw	Richmond	...	...	...	78.2	88.9	62.6	96	54	42	1.93	10	11	9	6	S.
AVERAGES	Western Maryland				70.4	79.9	61.3	....	....	..	39.6	2.81	13	12	5	9	E.
	Northern-Cent'l Md.				71.6	80.4	62.6	....	....	..	39.1	2.84	12.2	9.0	8.8	9.3	N. W.
	Southern Maryland.				74.3	84.2	64.4	....	....	..	42.8	1.82	14.0	9.0	7.0	6.2	S. W.
	Eastern Md. and Del.				72.8	81.6	63.4	....	....	..	44.1	1.60	10.9	12.7	6.4	5.2	N. E.
	Entire territory				72.3	81.5	62.9	....	....	..	41.2	2.22	12.5	10.7	6.8	7.4	N. E.

NOTE.—Letters of the alphabet are used to indicate the number of days that are missing from record : e. g. a—one day, b—two days, etc. (1)—H. SHRIVER. (2)—E. T. SHRIVER. *From tri-daily readings. † From bi-daily readings. ‡ Omitted in computing averages.

MONTHLY SUMMARY OF REPORTS FOR JULY, 1893.

STATIONS.	COUNTIES.	Altitude above sea in ft.	Latitude.	Longitude.	TEMPERATURE. Monthly Mean.	Mean of Max.	Mean of Min.	Degrees, Max.	Degrees, Min.	Monthly Mean.	Total Precipitation.	Clear Days.	Fair Days.	Cloudy Days	Rainy Days (.01 inch or more.)	Prevailing Wind.
WESTERN MARYLAND.																
Boettcherville	Allegany	....	39°33'	78°48'	75.6	86.7	64.6	99 54	45 1.50						6	N.W.
Cumberland (1).....	Allegany	650	39 39	78 46	77.7	89.1	67.1	97 57	40 1.38						7	....
Cumberland (2).....	Allegany	650	39 39	78 45	75.2	85.1	64.5	94 55	39 1.40	24	5	2	7	....	
Oakland	Garrett	2376	39 24	79 29	67.8	78.6	....	87 47	40 3.08	21	9	1	11	8. W.	
Sunny Side.........	Garrett	....	39 20	79 28	68.9	81.7	56.1	90 43	47 2.61	16	6	9	8	8. W.	
NORTHERN-CENTRAL MD.																
Baltimore		179	39 17	76 38	76.8	86.3	67.6	96 58	38 1.88	13	14	4	11	8. W.	
Darlington	Harford	300	36 47	76 14	74.4	84.6	64.2	95 55	40 2.98	28	1	2	6	W.	
Dist. Res., D. C....	....	....	39 9	77 0	*77.8	....	....	94 65	31 1.47	....	....	....	3	....	
Fallston	Harford	450	39 31	76 24	*73.4	....	....	96 57	38 2.68	16	10	5	7	8. W.	
Fenby	Carroll	950	39 33	77 5	*74.4	....	....	94 80	34 1.40	b19	b10	b6	6	N.W.	
Frederick	Frederick	280	39 24	77 18	77.2	88.4	66.0	97 55	42 1.80	17	14	0	8	N. W.	
Great Falls........	Montgomery .	....	39 0	77 14	*77.4	....	....	96 61	35 2.10	....	....	....	....	....	
McDonogh	Baltimore	535	39 23	76 44	74.8	b83.6	b66.3	92 56	36 1.94	....	....	....	3	....	
Mt. St. Mary's.....	Frederick	729	39 41	77 21	79.2	87.2	71.3	99 60	39 3.58	18	4	9	5	N. W.	
New Market........	Frederick	500	39 23	77 18	*75.4	....	....	95 56	39 1.70	23	8	0	4	N. W.	
Rec. Res , D. C....	....	....	38 52	77 0	*77.4	....	....	94 64	31 2.41	....	....	....	....	....	
Wa-hington, D. C..	....	112	38 53	77 0	77.0	87.1	66.9	97 57	40 1 44	17	11	3	11	S.	
SOUTHERN MARYLAND.																
Benedict...........	Charles	....	38 31	76 39	78.6	89.1	68.2	99 60	39 4.17	18	5	8	7	8. W.	
Charlotte Hall	St. Mary's ...	....	38 30	78 44	....	....	....	....	3.66	....	....	....	....	....	
Jewell	Anne Arundel	....	39 44	76 36	†78.0	....	....	....	2.50	28	3	0	5	8. W.	
Leonardtown	St. Mary's ...	....	38 18	76 40	78.2	87.5	69.2	95 57	38 3.64	17	14	0	7	8. W.	
Solomon's	Calvert	20	38 19	76 27	78.2	86.8	69.5	94 61	33 4.14	10	9	12	12	8. W.	
Upper Marlboro ...	Pr. George's..	....	38 47	76 45	77.4	88.7	65.8	99 55	44 2.54	20	8	3	6	N.W.	
EASTERN MD. AND DELAWARE.																
Barron Ck. Springs.	Wicomico ...	25	38 29	75 39	75.6	84.9	66.4	94 53	41 2.20	15	13	3	8	8. W.	
Cambridge	Dorchester....	....	39 39	76 46	81.3	87.9	74.7	97 67	30 7.12	0	26	5	9	N.W.	
Denton	Caroline	72	38 52	75 41	78.2	88.9	67.4	102 50	34 4.92	26	3	2	5	W.	
Dover, Del	Kent	....	38 42	75 31	75.3	83.8	66.7	94 56	38 5.29	24	3	4	12	8. W.	
Easton	Talbot	35	38 42	76 6	77.2	86.0	68.3	96 58	37 4.29	29	0	2	8	8. W.	
Kirkwood, Del.....	New Castle...	....	39 35	75 40	†74.8	....	....	....	....	....	....	....	....	
Milford, Del........	Kent	....	38 56	75 25	76.9	85.8	66.1	94 64	40 2.19	30	0	1	6	8. W.	
Millsboro, Del......	Sussex	....	39 41	75 15	75.8	86.4	65.1	97 51	46 2.47	21	7	3	6	N.	
Seaford, Del........	Sussex	....	38 40	75 35	75.6	85.9	65.1	96 56	40 6.08	....	....	....	7	....	
‡ VIRGINIA.																
Cape Charles......	Northampton	....	....	....	76.0	85.8	66.3	98 50	48 3.81	4	12	14	7	S. E.	
Norfolk...........	Norfolk......	....	....	....	79.0	88.0	70.0	95 62	33 6.11	16	13	2	11	S. W.	
Warsaw	Richmond	....	....	....	77.3	88.3	66.3	96 58	38 6.6	14	13	4	9	S.	
AVERAGES	Western Maryland	....	....	....	73.0	84.2	63.1	....	42.2 1.99	20.3	6.6	4	7.8	S. W.	
	Northern-Cent'l Md	....	....	....	76.3	86.2	67.0	....	36.9 2.12	18.1	9	2.9	6	N.W.	
	Southern Maryla d....	....	....	....	80.1	88.0	68.2	....	38.5 3.44	18.6	7.8	4.6	7.4	8. W.	
	Eastern Md. and Del..	....	....	....	76.6	86.2	67.5	....	40.5 4.33	20.7	7.4	2.9	7.8	8. W.	
	Entire territory	....	....	....	76.5	86.2	66.4	....	39.5 2.97	19.7	7.7	3.8	7.4	8. W.	

NOTE.—Letters of the alphabet are used to indicate the number of days that are missing from record : e. g. a—one day, b—two days, etc. (1)—H. SHRIVER. (2)—E. T. SHRIVER. * From tri-daily readings. † From bi-daily readings. ‡ Omitted in computing averages.

MONTHLY SUMMARY OF REPORTS FOR AUGUST, 1893.

STATIONS.	COUNTIES.	Altitude above sea in ft.	Latitude.	Longitude.	TEMPERATURE. Monthly Mean.	Mean of Max.	Mean of Min.	Degrees, Max.	Degrees, Min.	Monthly Range.	Total Precipitation.	Clear Days.	Fair Days.	Cloudy Days.	Rainy Days. (.01 inch or more.)	Prevailing Wind.
WESTERN MARYLAND.																
Boettcherville......	Allegany		39°33′	78°48′	*71.7	....	....	*97	*44	53	4.20	....	....	....	....	
Cumberland (1)...	Allegany	850	39 34	78 48	74.0	86.7	61.2	98	52	44	4.08	....	....	....	....	
Cumberland (2)......	Allegany	650	39 39	78 45	72.2	83.0	61.3	94	50	44	3.74	26	2	3	5	....
Oakland............	Garrett	2378	39 24	79 20	*63.6	....	....	*83	*44	39	3.71	12	17	2	7	S. W.
Sunny Side..	Garrett		39 20	79 28	*64.4	....	....	*90	*49	41	3.68	19	3	9	8	N. W. / S. W.
NORTHERN-CENTRAL MD.																
Baltimore............	....	179	39 17	78 36	74.6	83.8	65.7	90	57	33	1.81	19	6	8	6	S. E.
Darlington...........	Harford.....	300	38 47	78 14	a72.5	83.0	62.0	90	54	36	3.84	....	....	....	....	
Dist. Res., D. C.....	....		39 9	77 0	175.6	....	....	190	158	32	1.85	....	....	....	....	
Fallston............	Harford.....	450	39 31	76 24	71.8	....	....	87	54	33	6.26	....	....	....	....	
Fenby	Carroll	960	39 33	77 5	*72.8	....	....	*94	*54	40	4.20	22	9	0.	5	N.W.
Glyndon............	Frederick...	280	39 24	77 41	71.6	82.2	60.8	94	49	45	2.68	21	8	4	4	S. W.
Great Falls.........	Montgomery ...		39 0	77 14	174.6	....	....	193	166	37	2.81	....	....	....	....	
McDonogh	Baltimore....		39 23	76 44	73.8	83.0	64.2	91	55	36	2.82	....	....	....	....	
New Market........	Frederick...	600	39 23	77 18	*72.4	....	....	*95	*54	41	2.00	24	6	1	1	N.W.
Rec. Res., D. C.....	....		38 52	77 0	174.8	....	....	190	158	32	2.86	....	....	....	....	
Washington, D. C...	....	112	38 52	77 0	74.6	84.9	64.4	96	55	40	2.32	17	9	5	9	S.
Woodstock	Howard.....	392	39 20	78 49	c72.4	83.0	61.0	94	51	43	2.88	18	8	5	4	N.W.
SOUTHERN MARYLAND.																
Benedict	Charles....		38 31	76 39	75.9	85.3	66.5	93	57	36	3.18	21	0	10	5	S. W.
Solomon's.....	Calvert........	20	38 19	78 27	77.4	86.1	68.7	93	62	31	3.07	10	7	14	11	S. E.
Upper Marlboro....	Pr. George's..		38 47	76 45	73.9	86.1	62.7	91	53	38	3.80	20	5	6	10	S. E.
EASTERN MD. AND DELAWARE.																
Barron Ck. Springs.	Wicomico.....	25	38 29	75 39	75.2	84.3	66.1	93	56	37	2.67	9	19	3	7	N. E.
Cambridge.........	Dorchester...		39 39	76 46	78.8	85.6	72.0	95	63	32	1.68	....	....	....	....	
Easton	Talbot.....	35	38 42	76 6	74.6	83.4	66.2	92	57	35	4.24	24	4	3	7	W. / W.
Dover, Del........	Kent..........		39 9	75 31	73.8	81.7	65.8	90	57	33	3.03	21	5	5	9	N.W. / S. W.
Kirkwood, Del.....	New Castle....		39 35	75 40	179.2	....	....	....	....	....	....	....	....	....	....	
Milford, Del.......	Kent........		38 56	75 25	73.4	83.6	63.3	90	57	33	3.61	21	2	8	5	N.W.
Millsboro, Del......	Sussex		38 41	75 15	74.1	84.1	64.1	94	54	40	4.00	16	9	6	5	N. E.
Seaford, Del.......	Sussex		38 40	75 35	74.4	84.5	64.3	95	56	39	2.64	....	....	....	....	
VIRGINIA.																
Birdsnest...........	Northampton.				*76.3	....	....	92	54	28	4.65	14	7	10	5	N. E.
Cape Charles........	Northampton.				*74.9	....	....	90	63	27	5.25	....	....	....	....	
Norfolk............	Norfolk....				77.0	83.0	70.0	91	82	29	5.71	8	13	10	13	N. E.
Warsaw	Richmond...				74.6	....	....	94	56	38	2.93	17	7	7	5	S.
AVERAGES	Western Maryland..				69.2	84.8	61.2				44.2	3.87	19.0	7.0	5.0	7.0	S. W.
	Northern-Cent'l Md..				73.4	83.4	63.0				37.7	3.01	20.2	7.3	3.5	3.8	N.W
	Southern Maryland..				76.7	86.5	65.9				35.0	3.34	17.0	4.0	10.0	8.7	S. E
	Eastern Md. and Del.				75.6	84.9	65.9				35.6	3.14	18.2	7.8	5.0	5.8	N. W
	Entire territory......				73.5	84.5	64.0				38.1	3.34	18.6	6.5	5.9	6.7	N.W

NOTE.— Letters of the alphabet are used to indicate the number of days that are missing from record : e. g. a—one day, b—two days, etc. (1)—H. SHRIVER. (2)—E. T. SHRIVER. * From tri-daily readings. † From bi-daily readings. ‡ Omitted in computing averages.

MONTHLY SUMMARY OF REPORTS FOR SEPTEMBER, 1893.

STATIONS.	COUNTIES.	Altitude above sea in ft.	Latitude.	Longitude.	TEMPERATURE. Monthly Mean.	Mean of Max.	Mean of Min.	Degrees, Max. Degrees, Min.	Monthly Range.	Total Precipitation.	Clear Days.	Fair Days.	Cloudy Days.	Rainy Days. (1.01 inch or more.)	Prevailing Wind.
WESTERN MARYLAND.															
Boettcherville	Allegany		39°33'	78°48'	63.1	.....		88 32		56 3.10	...	...	...		
Cumberland (1)	Allegany	700	39 39	78 46	a66.0	74.1	58.0	87 28		49 1.99	...	...	...		
Cumberland (2)	Allegany	700	39 39	78 45	64.0	72.4	55.6	86 35		51 1.97	16	6	5	3	...
Oakland	Garrett	2380	39 25	79 20	*58.0			80 31		49 1.84	17	8	5	10	S. W.
Sunny Side	Garrett		39 20	79 28	e60.6	71.6	49.6	82 28		54 1.40	12	9	9	7	S. W.
NORTHERN-CENTRAL MD.															
Bachman's Valley	Carroll		39 40	77 01	*60.2	.....		82 33		49 1.80	18	5	7		S. N. W
Baltimore		179	39 17	76 37	66.6	74.3	58.8	88 44		44 1.80	12	9	9	8	W.
Darlington	Harford	309	39 47	76 14	64.0	74.2	53.7	84 40		44 2.98	19	5	6	5	N. W.
Dist. Res., D. C.			39 9	77 0	†65.8	.....		83 40		43 3.96	...	...	...		
Fallston	Harford	300	39 30	76 24	*63.0	.....		86 41		45 2.84	...	...	...		
Fenby	Carroll	950	39 33	77 5	*63.2	.....		82 40		42 2.70	10	14	6	7	N. W.
‡ Frederick	Frederick	400	39 24	77 24	65.6	73.9	57.2	86 37		49 2.09	12	11	7	11	....
Glyndon	Baltimore	280	39 2	77 14	63.5	70.3	56.7	84 41		43 2.63	15	6	9	8	N. W.
Great Falls	Montgomery		39 0	77 14	†65.2	.....		85 39		46 2.27	...	...	...		
McDonogh	Baltimore	545	39 23	76 46	c66.2	73.6	58.9	84 48		36 2.01	...	...	...		
Mt. St. Mary's	Frederick	715	39 43	77 20	66 2	73.9	58.6	88 43		45 2.59	9	13	8	10	W.
New Market	Frederick	500	39 10	77 15	*64.0	.....		89 38		51 1.48	17	5	8	3	N. W.
Rec. Res., D. C.			38 52	77 0	†60.0	.....		82 41		41 3.18	...	...	...		
Washington, D. C.		112	38 53	77 0	66.0	74.5	57.4	88 42		46 3.91	13	7	10	9	N. W.
Woodstock College	Baltimore	400	39 19	76 51	64.0	73.0	55.0	86 37		49 2.03	15	10	5	3	N. W.
SOUTHERN MARYLAND.															
Benedict	Charles		38 31	76 39	68.2	77.9	58.5	91 42		49 1.41	19	0	11	4	W.
Solomon's	Calvert	20	38 19	76 27	70.0	78.7	61.6	90 46		60 2.56	8	7	15	9	S. E.
Upper Marlboro	Pr. George's		38 47	76 45	a65.5	75.5	55.3	89 39		50 4.42	18	5	7	8	N. N. W.
EASTERN MD. AND DELAWARE.															
Barron Ck. Springs	Wicomico	25	38 30	75 39	67.2	75.1	59.2	88 43		45 3.61	11	8	11	7	N. W.
Cambridge	Dorchester		38 39	76 7	71.2	78.3	64.0	93 50		43 2.49	0	8	8	8	S.
Easton	Talbot	35	38 42	76 6	67.5	75.4	59.6	88 44		44 2.10	18	3	9	4	N. W.
Dover, Del	Kent	40	39 10	75 30	65.8	73.7	58.0	87 44		43 3.69	15	8	6	11	S. W
Kirkwood, Del	New Castle		39 33	75 41	†65.9	.....				...	...	...	...		
‡ Milford, Del	Kent		38 56	75 25	67.1	75.2	59.0	84 44		40 4.32	18	3	9	5	W
Millsboro, Del	Sussex		38 41	75 15	66.2	75.6	56.8	88 43		45 6.17	12	10	8	10	N.
Seaford, Del	Sussex		38 40	75 35	67.4	77.3	57.6	88 40		48 3.76	...	...	...		
‡ VIRGINIA.															
Birdsnest	Northampton				69.5	.....		88 49		39 7.75	9	12	9	13	N. E.
Cape Charles	Northampton				66.9	.....		84 49		35 5.46	12	6	12	9	....
Norfolk					71.0	78.0	64.0	89 51		38 6.29	14	8	8	9	S. W.
Warsaw	Richmond				67.1	76.7	57.5	92 40		52 4.48	15	8	7	5	S.
AVERAGES {	Western Maryland				62.3	72.7	54.4			51.8 2.06	15.0	7.7	6.3	6.7	S. W.
	Northern-Cent'l Md				64.1	73.4	57.0			44.6 2.54	14.2	8.2	7.5	5.9	N. W.
	Southern Maryland				67.9	77.4	58.0			53.0 2.80	15.0	4.0	11.0	7.0	N. W. W.
	Eastern Maryland				67.0	75.1	57.5			44.7 3.64	15.6	5.8	8.4	8.2	S. E. N. W.
	Entire territory				65.3	74.8	56.7			48.5 2.76	15.0	6.4	8.3	7.0	N. W.

NOTE.—Letters of the alphabet are used to indicate the number of days that are missing from record: e. g. a—one day, b—two days, etc. (1)—H. SHRIVER. (2)—E. T. SHRIVER. * From tri-daily readings. † From bi-daily readings. ‡ Omitted in computing averages.

MARYLAND STATE WEATHER SERVICE.

MONTHLY SUMMARY OF REPORTS FOR OCTOBER, 1893.

STATIONS.	COUNTIES.	Altitude above sea in ft.	Latitude	Longitude	TEMPERATURE.						Total Precipitation.	Clear Days.	Fair Days.	Cloudy Days.	Rainy Days. (.01 inch or more.)	Prevailing Wind.
					Monthly Mean.	Mean of Max.	Mean of Min.	Degrees, Max.	Degrees, Min.	Monthly Mean.						
WESTERN MARYLAND.																
Boettcherville	Allegany	....	39°33'	78°48'	52.8	....	....	80	22	58	4.70	....	....	....	....	
Cumberland (2)	Allegany	700	39 39	78 45	54.0	63.2	44.6	80	25	55	4.37	16	7	8	7	
Oakland	Garrett	2380	39 26	79 20	48.4	59.0	37.7	74	18	58	5.00	13	13	5	12	
Sunny Side	Garrett	....	39 20	79 28	49.0	61.1	36.9	81	13	68	5.02	....	....	....	....	S. E.
NORTHERN-CENTRAL MD.																
Bachman's Valley	Carroll	....	39 40	77 01	*49.4	....	....	*75	*20	56	4.45	20	4	7	4	N.W.
Baltimore		179	39 17	76 37	57.0	64.8	49.2	84	31	53	3.44	18	4	9	9	N.W.
Darlington	Harford	300	39 47	76 14	55.3	65 4	45.2	78	26	53	2.59	....	....	....	....	
Dist. Res., D. C		39	9	77 0	155.8	....	....	175	128	48	3.30	....	....	....	....	
Fallston	Harford	300	39 30	76 24	*54.7	....	....	*80	*30	50	5.15	....	....	....	....	N.
Fenby	Carroll	950	39 33	77 5	*57.5	....	....	*78	*30	48	5.30	....	....	....	....	
‡ Frederick	Frederick	400	39 24	77 24	55.9	64.9	45.9	81	28	53	3.78	21	5	5	6	
Glyndon	Baltimore	280	39 24	77 41	55.7	63.3	49.1	79	30	49	3.93	17	7	7	5	S. E.
Great Falls	Montgomery	....	39 0	77 14	155.2	....	....	177	128	49	4.29	....	....	....	....	
McDonogh	Baltimore	545	39 23	76 46	b56.8	65.6	47.7	78	32	46	3.58	....	....	....	....	W.
Mt. St. Mary's	Frederick	720	39 43	77 20	54.4	64.8	43.9	84	22	62	4.43	15	8	8	8	W.
New Market	Frederick	500	39 10	77 15	*53.9	....	....	*79	*25	54	3.20	21	2	8	8	N.W.
Rec. Res., D. C		....	38 52	77 0	155.7	....	....	177	129	48	4.36	....	....	....	....	
‡ Tanoytown	Carroll	....	39 40	77 9	....	....	....	....	....		3.25	....	....	....	....	
Washington, D. C		112	38 52	77 0	58.5	55.9	47.1	83	26	57	4.11	15	8	8	7	N.W.
Woodstock	Howard	400	39 19	76 51	53.9	64.3	43.5	81	24	57	5.60	....	....	....	....	
SOUTHERN MARYLAND.																
Benedict	Charles	....	38 31	76 39	58.7	56.9	44.5	83	31	52	3.09	28	0	5	4	E.
Charlotte Hall	St. Mary's	....	38 30	76 44	c56.6	66.3	46.9	77	29	48	4.48	....	....	....	....	N. E.
Solomon's	Calvert	20	38 19	76 27	60.4	69.0	51.9	88	36	47	4.88	15	5	11	7	S. E.
Upper Marlboro	Pr. George's	....	38 47	76 45	56.0	66.4	45.0	82	26	55	5.45	....	....	....	....	N. E.
Valley Lee	St. Mary's	....	38 12	76 35	159.3	....	....	....	....		6.40	23	0	8	5	N.W.
EASTERN MD. AND DELAWARE.																
Barron Ck. Springs	Wicomico	25	38 30	75 39	58.8	66.4	47.3	85	26	59	2.85	18	8	5	11	N.W. S. E.
Cambridge	Dorchester	....	38 39	76 7	61.9	68.8	55.0	88	36	52	3.61	....	....	....	....	
Dover, Del	Kent	40	39 10	75 30	56.8	65.4	48.1	80	31	49	4.37	21	3	7	10	E. S. E.
Easton	Talbot	35	38 42	76 6	58.4	67.5	49.3	82	31	51	4.04	22	3	6	6	N.W.
Kirkwood, Del	New Castle	....	39 35	75 41	156.2	....	....	....	....	....	....	....	....	....	....	
Milford, Del	Kent	29	38 45	75 25	57.4	66.3	48.6	82	30	52	3.87	22	5	4	5	N.W.
Millsboro, Del	Sussex	....	38 41	75 15	56.8	66.7	48.4	81	29	52	2.88	21	5	5	6	N.
Seaford, Del	Sussex	....	38 40	75 35	54.0	65.0	43.0	80	28	52	3.16	....	....	....	....	
‡ VIRGINIA.																
Birdsnest	Northampton	....	....	....	*60.5	....	....	*83	*36	49	4.75	17	5	9	5	N. E.
Cape Charles	Northampton	....	....	....	60.9	70.1	51.7	78	29	49	3.22	....	....	....	....	
Norfolk		....	....	....	62.0	68.0	55.0	83	40	43	2.85	17	6	8	8	N. E.
Warsaw	Richmond	....	....	....	57.0	66.7	47.3	80	28	52	6.13	22	5	4	5	R.
AVERAGES {	Western Maryland	....	....	....	51.0	61.1	39.8	....	....	59.8	4.77	14.5	10.0	6.5	9.5	S. R.
	Northern-Cent'l Md.	....	....	....	55.1	64.7	46.4	....	....	52.1	4.29	17.7	5.5	7.8	6.0	N. W.
	Southern Maryland	....	....	....	58.2	67.6	48.2	....	....	50.8	4.85	21.3	1.7	8.0	5.7	S. E.
	Eastern Md. and Del.	....	....	....	57.3	66.9	48.2	....	....	52.4	3.74	20.6	4.6	5.4	8.0	N. W.
	Entire territory	....	....	....	54.9	65.3	45.5	....	....	53.8	4.42	18.6	3.2	6.9	7.3	N W. S. W.

NOTE.—Letters of the alphabet are used to indicate the number of days that are missing from record: e. g. a—one day, b—two days, etc. (1)—H. SHRIVER. (2)—E. T. SHRIVER. * From tri-daily readings. † From bi-daily readings. ‡ Omitted in computing averages.

MONTHLY SUMMARY OF REPORTS FOR NOVEMBER, 1893.

STATIONS.	COUNTIES.	Altitude above sea in ft.	Latitude	Longitude	Monthly Mean	Mean of Max.	Mean of Min.	Degree. Max.	Degree. Min.	Monthly Range.	Total Precipitation.	Clear Days.	Fair Days.	Cloudy Days.	Rainy Days. (.01 inch or more)	Prevailing Wind.
WESTERN MARYLAND.																
Sunny Side	Garrett		39 20	79 28	34.1	42.3	25.9	66	0	66	2.55	12	6	12		9 N. E.
Oakland	Garrett	2380	39 25	79 20	35.2	44.5	25.9	65	1	59	3.44	11	4	15		10 ....
Boettcherville	Allegany		39 33	78 48	*38.6	....	....	66	12	54	2.90	....	....	....		....
Cumberland (1)	Allegany	700	39 39	78 46	b44.6	b52.0	b37.1	65	20	45	2.80	9	7	14		6 N.W.
Cumberland (2)	Allegany	700	39 39	78 45	39.0	31.4	31.4	62	14	48	2.01	15	4	11		5 ....
NORTHERN-CENTRAL MD.																
Mt. St. Mary's	Frederick	720	39 43	77 20	39.0	48.8	29.3	60	13	47	4.57	14	5	11		6 W.
Frederick	Frederick	400	39 24	77 24	41.7	49.0	34.4	62	18	43	3.06	11	12	7		8 ....
New Market	Frederick	500	39 10	77 15	*42.1	....	....	61	16	45	2.79	15	7	8		4 N.W.
Taneytown	Carroll		39 40	77 9	....	....	....				3.49	25	0	5		5 S.
Bachman's Valley	Carroll		39 40	77 1	*34.7	....	....	52	16	42	4.64	14	5	11		8 N.W.
Fenby	Carroll	950	39 33	77 5	*40.8	....	....	60	18	42	5.90	11	13	6		6 N.W.
McDonogh	Baltimore	545	39 22	76 46	45.2	52.9	33.4	64	19	45	2.69	....	....	....		....
Woodstock College	Baltimore	460	39 19	76 51	41.0	51.6	30.5	62	15	47	5.70	6	7	17		6 N.
Baltimore		170	39 17	76 37	45.6	50.3	36.6	62	22	40	3.52	15	6	9		11 N.
Fallston	Harford	300	39 30	76 24	*40.9	....	....	60	19	41	4.53	....	....	....		....
Darlington	Harford	300	39 47	76 14	41.4	50.3	32.6	62	19	43	3.87	19	2	9		5 N.W.
Great Falls	Montgomery		39 0	77 14	†42.0	....	....	64	15	49	3.58	....	....	....		....
Dist. Res., D. C.			39 9	77 0	†43.3	....	....	64	20	44	4.00	....	....	....		....
Rec. Res., D. C.			38 52	77 0	†45.2	....	....	64	20	44	4.20	....	....	....		....
Washington, D. C.		112	38 52	77 0	43.4	51.6	35.4	66	21	45	4.30	13	6	11		15 N.W.
SOUTHERN MARYLAND.																
Upper Marlboro	Pr. George's		38 47	76 45	42.6	52.4	32.9	68	17	51	4.42	14	11	5		9 N. E.
Benedict	Charles		38 31	76 39	45.1	53.8	36.9	70	21	49	2.86	13	3	14		5 S. W.
Charlotte Hall	St. Mary's		38 30	76 44	a44.6	55.4	35.9	73	18	55	....	15	9	5		....
Valley Lee	St. Mary's		38 12	76 35	†46.9	....	....				8.27	....	....	....		N. E.
Solomon's	Calvert		38 19	76 27	47.2	55.1	38.4	70	24	46	4.66	7	10	13		7 N.W.
EASTERN MD. AND DELAWARE.																
Chestertown	Kent		39 13	76 7	b48.4	48.0	34.9	69	22	38	2.40	....	....	....		....
Cambridge	Dorchester		38 39	76 7	48.2	53.2	43.1	67	30	37	6.25	25	0	5		5 N.W.
Denton	Caroline		38 47	75 41	46.8	58.3	35.3	71	21	50	2.55	8	11	11		4
Barron Ck. Springs	Wicomico	25	38 30	75 39	45.2	54.3	35.8	73	18	55	3.24	10	14	6		8 N. E.
Kirkwood, Del	New Castle		39 35	75 41	†42.2	....	....				....	....	....	....		....
Dover, Del	Kent	40	39 10	75 30	44.6	52.1	37.0	65	22	43	2.80	16	8	6		11 N.W.
Milford, Del	Kent	20	38 45	75 25	45.0	53.8	36.5	69	21	48	3.08	19	5	6		4 N.W.
Seaford, Del	Sussex		38 40	75 35	44.9	55.3	34.5	70	20	50	3.15	....	....	....		....
Millsboro, Del	Sussex		38 41	75 15	44.7	54.5	34.9	71	18	53	3.41	18	8	4		9 N. E.
‡ VIRGINIA.																
Birdsnest	Northampton				*49.3	....	....	76	25	51	7.29	10	7	13		5 N. E.
Cape Charles	Northampton				*48.6	....	....	69	27	42	8.24	....	....	....		....
Norfolk					50.3	56.8	43.8	74	25	49	6.75	10	10	10		11 N. E.
Warsaw					45.5	54.8	36.2	73	19	54	3.42	13	11	6		4 N.
AVERAGES	Western Maryland				38.8	42.6	30.1	....	54.4	2.82	11.8	5.2	14.2		7.5 N. E. N.W.
	Northern-Cent'l Md				41.4	50.6	33.1	....	44.1	3.72	14.3	6.3	9.4		7.4 N.W.
	Southern Maryland				45.3	54.6	38.3	....	50.1	5.05	11.5	8.2	9.2		7.2 N. E. N.W.
	East. Md. and Del				45.3	53.7	36.5	....	46.8	3.37	16.0	7.7	6.3		6.8 N.W.
	Entire territory				42.6	50.4	34.5	....	48.8	3.74	13.4	6.8	9.8		7.2 N.W.

NOTE.—Letters of the alphabet are used to indicate the number of days that are missing from record: e. g. a—one day, b—two days, etc. (1)—H. SHRIVER, (2)—E. T. SHRIVER. * From tri-daily readings. † From bi-daily readings. ‡ Omitted in computing averages.

DAILY PRECIPITATION FOR APRIL, 1892.

Stations.	1	2	3	4	5	6	7	8	9	10	11	12	13	14	15	16	17	18	19	20	21	22	23	24	25	26	27	28	29	30	31	Total.
Baltimore	.02	T			.21	T	T	.06	.06					.34	.42		T	.50		.09	.62	.52	.06		T	T			.25			3.15
Barron Crk.Spr.	T	T			.40	T		.79	.01					1.20	.06			.33				.34							.89	.51		6.68
Boettcherville							1.10			.05				.60	.29			.60		.10	.70	.70	.10						.10			3.50
Charlotte Hall	.25													.10	.80			.75	.10	.16	1.60	.75							.67			4.77
Cumberland (1)					.56									.84	.02			.54	.10	.43	.26	.73	.04						.04			3.51
Cumberland (2)					.49			.06						.84	.02			.53				.60							.04			3.21
Darlington					.22	.05		.08	.08					.64				.25	.65		.61								.18			2.05
Dist. Res., D. C.	.25		.05				.08								1.02	.16		.17			.86	1.01	1.10		.08				.18	.51		5.29
Dover, Del									.09						.85			.57		.16	.61		.04						.62			4.08
Easton	.02	T	T		.21		.02	.20	.09					.83	.10	.16		.76		.18		1.44	.25						.70			4.61
Fallston	.02	.01			.22	.27	.53	.03	.04					.15	.50	T		.53	.37		.47	.52	.63						.11	.30		2.89
Frederick	.50	T			.15			.12						.40	.45	.10		.20	.70	.12	.63	.20	.02						.10	.40		2.36
Great Falls	.01					.27			.01						1.33			.61			1.84	1.84			T				.77			4.80
Leonardtown	T				.15				.03					.90				.54	.73	.16	.03	.38			.05				.18			5.08
McDonogh	.01	.02			.32			.20	.02						.90			.56		.12	.08	.36							.77			2.03
Mt. St. Mary's	.02	T	.04		.32			.12							.72			.54		.42	.16	.79							.18	.40		2.80
New Market	T	T			.40		T	.10	.01						.65	.10	.05	.56			.87	1.12							.32			3.06
Rec. Res., D. C.	.63		.08			.12	.07	.10	.03						.94			.14	.73		.87	.07	.96		T				.04	.56		5.84
Seaford, Del		T						.45	.04						1.11		.10	.70				2.54							.67			5.58
Solomon's								.23						1.18				.57			.65	1.16			T				.75			5.23
Taneytown	.03		T		.40	T		.07	.02	T				.40	.37			.67	.30	.02	.56	.41	.08		.11				.36			2.96
Wash'gt'n, D. C.	.02	.04			.03	T	.07	.06	.02					.80	.17			.61	.30	.17	.72	.87			T		T		.65	.40		4.52
Woods'k College					.30		T	.02	T					.73	.05			.50		.61	.61	.30			T	T			.25			3.02
Norfolk, Va.							.65	1.81	T					.10	.28			.71	T	.49	.21	.82	1.59		.09	.03			.50			6.96

DAILY PRECIPITATION FOR MAY, 1892.

STATIONS.	1	2	3	4	5	6	7	8	9	10	11	12	13	14	15	16	17	18	19	20	21	22	23	24	25	26	27	28	29	30	31	Total
Baltimore.		.01		.05	.05					.08	.52			44	1.83	.02		.05	1.53	T	.42	T	.04				.72	.46	.03	T		6.36
Barren Crk.Spr.		.13		.06							1.20			.36					.34		.33	.62	.33				.16		T			3.44
Boettcherville.		.40		.10		.30				.40	.20			.90	.30			1.00	1.00	.46	.50		.20		.20		.10	.20	.29	.40		4.70
Cumberland (B.		.16				.19				.43			.21	.12	.21		.65			.17					.20		.01	.20				3.81
Denton						.70				1.00	.76			1.00					1.19								.33					3.89
Dist. Res., D. C.			.30								.22	.36		.36	.45				.34													4.82
Dover, Del.		.08			.06	.12					1.17			.02	1.77					.37		.21	.03				.04					6.69
Easton		.05				.23				.06	1.64			.33	.43			1.25	.25		*	.11	.31		*		.43		.13			5.65
Fallston		.10				.08					.43			1.15				16	2.28	.25	.40		.02		1.31		.12	1.31				6.10
Frederick.		.06			T	.05					.13			.26	.08	.22			.01	*	.65	.13	.04				.07	.58			.30	2.16
Great Falls			.14								.26	.64		.23	.40					.25	.20	.75	T				.06	1.06	.16			3.74
Jewell		T		T		T					1.35		*	.75	.27				.50								T					4.75
McDonogh		.30			.06	.07				.16	.29	.36		.26		.42	T	.17	.85	.28	.80	.13					.60	.80	.07	.21		3.92
Mt. St. Mary's.		.22			.06	.17				.27	.08			.98					.73	*	.68						.30	.18	.18	.01		3.17
New Market		.13				.32	.33			*				.43	.76			1.36	1.36	.47		.24				1.27	.31	1.27				5.55
Roc. Res., D. C.			.09								.23	.06		.21	.38				.24		.14	.50	.09				.09		.09			4.81
Seaford, Del.		.21				.31					1.15	T		.29					.34		.40	.18	.36		T		.19		T			3.16
Solomon's		T			T	.21				.06	1.38		.06	.21		.09			1.80	.70				1.12	.03		.36	.54	T	.21		2.99
Taneytown		.16				.04				.02	.10			.16	.76	.42	T	.21	.33	.04			.92					T	T			5.37
Wash'gt'n, D. C.		.07	T			.07	T			.10	.47			.58	.01				.77	*	.39	.03	.02			1.20	.02	1.20	T		.10	4.07
Woods'k College											.30			1.34	.64				.24		1.67	.50	.04				.05		T			4.78
Norfolk, Va.							.83			T	.64														T						T	3.76

DAILY PRECIPITATION FOR JUNE, 1892.

STATIONS.	1	2	3	4	5	6	6	7	8	9	10	11	12	13	14	15	16	17	18	19	20	21	22	23	24	25	26	27	28	29	30	31	To-tal.
Baltimore				.01	.27				.57	.37	T					T		.07	.11		.38	.06	T	.46		.34		1.58	.02		.63		4.87
Barren Ck. Spr.			.90	2.50		.30			.11	.03	T								.45		.04							.29			.60		1.82
Boettcherville	.20		.90		.11			•	.80	.70	.30						.50	.04	.20	.30	.20	.02		T		.12		.10		.10			8.60
Cumberland (1)	.75		2.55	4.64				1.00	1.00	.65	.40						.53	.60	.45	.30							.07	.10			.31		7.31
Cumberland (2)	.40		.68	.28		.08		.55	.16	.62	.38				.04		.67	.68	.97	.27		.02			.04			.08			.22		10.80
Darlington				.28					.16	.12	.07	.02						.15		.05	.05				.30	.12		3.34	.57		.05		4.02
Dist. Res., D. C.					.17	.02			•		.58					T				.20	.03	.29						.02			.48		1.42
Dover, Del					.08					.45	.30							.38						T			.70	1.20			.50		1.45
Easton									.28	.46	.30									.08	.08	.02		.93	.33	.23		1.57			.05		3.06
Fallston	.02				.12	.05		.40	.40	.24	.11	.56			.33			.23	.28	.36	.06	.02	.10			.23		.19		.08	.19		3.35
Frederick				.23		.12	T			.63	.75								.33	.12								.80			.87		2.80
Great Falls			.80		.65				2.50	.63	.65								.66	.04			.87		•			.82	.86				2.10
Jewell	.87	.85		T				.10	.40	.28	.15			T			.43	.46	T	.69	.55		.10		.84	.84		.45	.59		.16		5.58
McDonogh				T	T			.20	.90	.80	1.09						.50	.50	T	.59	.05			.93				.05			1.05		2.90
Mt. St. Mary's				T	T				.01	.55				T				.14	.02	.83	.06						T	.20			.78		4.52
New Market					.16			.08	.06	.12	T								T	.22	.03				.01			1.10			.01		3.46
Rec. Res., D. C.		.35	.88		.40			T	T	.90	T							.50	T	.90	.14			.41				.53	.02		.06		3.01
Seaford, Del								.08	T	.09																		.40	.50		.30		2.34
Solomon's												.05						.48		.14						.08	.01	T		.15	.18		4.00
Taneytown																																	3.03
Wash'gt'n, D. C.																																	2.59
Wood'k College																																	2.90
Norfolk, Va.																																	4.83

NOTE.—"T" indicates a trace of rain or snow.   •Amount included in next measurement.

DAILY PRECIPITATION FOR JULY, 1892.

STATIONS.	1	2	3	4	5	6	7	8	9	10	11	12	13	14	15	16	17	18	19	20	21	22	23	24	25	26	27	28	29	30	31	Total
Baltimore	.03		.91			T					.20			.54	T			.43		.43		T	.42				T	.82		.16	.04	4.07
Barren Crk.Spr.	.09		.16	T							.44			.25		.05		.55		.55		.57	.10				*	.34		.12	.78	8.97
Boettcherville			.70								T				T	.10													T	.20		1.10
Cumberland (1)			.62									.06						.25		.30		.06	.10				.16		.10			1.22
Cumberland (2)			.09								.10	.35		.54	.00			.30		T	.80	.07					.02		.10			1.16
Darlington	.48	2.15		.44							.03			.61	.27	.00		T		1.11			.23			T	.21	1.88		.06	.70	4.63
Dist. Res., D. C.	.75	.07	.27								.08			.06				.41												2.87	.29	6.40
Dover, Del	.26										.29			.12				1.18		.01							.04			.36	.11	4.35
Easton	T										.10			.75				.33				.18					.44		T	.10	2.63
Fallston	.09	.10	.07	2.10							.09	T						.00				.45	.45				.75				.06	6.06
Frederick			*1.12								.04			.04	.06			.00				T					.08	.04			4.11	2.29
Great Falls	1.54	.06	1.63								T			1.56				2.00		.00										.15	.06	4.02
Jewell	T		.12				.71					.08	.64		.16							.54					.90			.08		4.38
Leonardtown				.28							.70	T		.25				.00		.01							.80				.07	4.67
McDonogh			1.00															.00			2.00						1.93				.18	5.84
Mt. St. Mary's	.04	.01	.40								.02	T		T	1.48			.00		.00		.00				T	.85			.17	.08	5.80
New Market	.00	1.70									.62							.07		.07		.43			T		.60			.18	.07	4.89
Res. Res., D. C.	.10	.04		.61							.56	.98		.17	.66	.30		.24		.33			.18				.20			.83		2.90
Seaford, Del	.25	.06	T								.58							.24		.57			T			T	.55				1.12	2.40
Solomon's	.07	.02	.06								.21			.12				.01									.48			.80		4.64
Tanaytown	1.20	.01	.80								.20							.03				.07	*1.25								.02	6.08
Washington, D. C.	.13	T	.31			.50					.18	T		1.44	.00			.86	.17				.82				.48	1.00	T	.18	.80	8.27
Norfolk, Va	1.06	.53				.06	T			.09	.50	1.54	T	.84				.10											.70		.55	

Note.—"T" indicates a trace of rain or snow.   * Amount included in next measurement.

DAILY PRECIPITATION FOR AUGUST, 1892.

STATIONS.	1	2	3	4	5	6	7	8	9	10	11	12	13	14	15	16	17	18	19	20	21	22	23	24	25	26	27	28	29	30	31	Total
Baltimore	.54		.03		.19					.04	.28	T									.08	.02	T	.04	.27	.04	T	T				1.83
Barron Crk.Spr.		.34			.97							T									.08	.78	.03		.06					.18	.07	2.40
Boettcherville											.10												.30							.05		1.00
Cumberland (1)	.81																			.40					.37							2.03
Cumberland (2)	.56																			.41					.30						.07	1.90
Darlington	.06	1.70								.07	.63									.14	.08				.70						.09	3.30
Dist. Res., D. C.	.49	.24		.46							.06										.08		.12		.21			.12			.07	1.40
Dover, Del		.67			.48				.69														.12								.07	2.54
Easton		.18			.43			.01			.01	.18											.12		.15			.01				1.09
Fallston		1.30			.56						.34										.40			.35	.73				.90			4.10
Frederick	*	.21		.35	.66					.02	.64													.12	.36						.08	1.68
Great Falls	.14				.30					.02														.34								.78
Jewell		1.12			3.35																.10											2.47
Leonardtown	.08		.53							T	T												.40	T	.54	.25					.02	1.31
McDonogh		1.12		.01						T	.28	T									.32		.01	1.13							.02	2.39
Mt. St. Mary's	T	.81			.05					1.12	.32	T												T							T	3.39
New Market		.05		1.30						T	.88													.06	.11						.14	1.27
Rec. Res., D. C.	.09	.40	.04									T											T	.06	T	.05					.15	1.19
Seaford, Del.	.06	.28		T	.34							T									.14	.11	.28	.32	T					1.29	.15	1.89
Solomon's	.06	.66			.05							T											.04	.01	T				.01		.08	2.89
Taneytown										T	.53	T									T		.04	T	.06	T	T					.58
Wasb'g'n, D. C.		.15			.84					T	T	T									T		T	T	.80	T					T	.85
Woods'&College		1.30			.85					T	T	T										.31	T	T							.08	2.13
Norfolk, Va.	.05	.05	.17		.04							T										.31	.33	.06					.05	.06	2.13	3.53

NOTE.—" T " indicates a trace of rain or snow.   *Amount included in next measurement.

DAILY PRECIPITATION FOR SEPTEMBER, 1892.

Stations.	1	2	3	4	5	6	7	8	9	10	11	12	13	14	15	16	17	18	19	20	21	22	23	24	25	26	27	28	29	30	31	Total.
Baltimore						.01			.24		T		.39	.83		.01					.09	.27	.34	T	.16							2.38
Barron Ck. Spr.					.29									.83								.38	.38		.64							2.06
Boettcherville													1.60									.50	.30		.30							2.90
Cumberland (1)					.55			.02	.04				.90	.90							.35	.38	.22	.05								2.25
Cumberland (2)					.52				.10				1.06	T								.20			.22	.30						2.36
Darlington						.13							1.20	1.75	.08							.51			1.08							2.77
Dist. Res. D. C.									.08					.76							.36	.59	.25	.06	.06							4.66
Dover, Del.						.03																.84	.84		1.08							2.71
Easton						.12			.07				.08	.65	.24							.19	.28		.21							1.84
Fallston					.18							T		1.99								.65			.50							3.29
Frederick					.36			.05	.04				.46	.14							T	.25	.41	.05								5.82
Great Falls					.57			.20					.64	.64	.08							.54	.62		.36							2.04
Jewel					.14			.12	.10				.90	T								1.80	1.80									3.08
Leonardtown						T			.07				1.4	1.31								.86	.86	.03								2.29
McDonogh					.64			.41						.06	.10						.02	.30	.42	.06								3.32
Mt. St. Mary's													4.38	.02							T	.37	.09									5.52
New Market					.57			.60	.05				1.70	.26								.70	.62									2.77
Roc. Res. D. C.														1.32							T	.51	2.00	.05	.87							4.10
Seaford, Del.								.04						.94											.25							1.81
Solomon's					.06			.30			T		.25	.43							.12	.13	.36	.10								1.57
Taneytown													1.60	.20							.18		.05									2.44
Wash'gt'n, D. C.					T	.13			.02			T	.47	.70	.01			T			.18	1.46	1.49									3.65
Wood'x College					.53			.05					2.11	.18						T	.31	.37										3.53
Norfolk, Va.	.18				.52							T	.18	.23				T		T	T	T	.03		.76							1.33

DAILY PRECIPITATION FOR OCTOBER, 1892.

STATIONS.	1	2	3	4	5	6	7	8	9	10	11	12	13	14	15	16	17	18	19	20	21	22	23	24	25	26	27	28	29	30	31	Total.
Baltimore				T	.05			.22	.02										T			T	T						.02			0.36
Barron Crk.Spr.					T			.04														.07	T						T			0.09
Boettcherville				.20																												0.20
Cumberland (1)				.20																		.07							.04			0.27
Cumberland (2)			.17						.29																							0.24
Darlington			.03						.31																							0.38
Dist. Res., D. C.									.33																							0.31
Dover, Del																							.03									0.46
Easton									.40														.03						.10			0.79
Fallston				T				.14											T		T	T							.08			0.45
Frederick				.04	.01				.10							.20													.05			0.19
Great Falls				T																		T		T								0.10
Jewell			T					.50														T						T				0.50
Leonardtown								.21	1.10							.03			T													1.13
McDonogh				.03				.18	.01							.09					.01	T							.02			0.34
Mt. St. Mary's								.21																								0.22
New Market									.26																							0.21
Rec. Res., D. C.	.26																															0.26
Seaford, Del			T					.40	T												T	T	.02	T					.19			0.84
Solomon's				.07				.50													T	T	T						.02			0.67
Tabeytown																																0.00
Wash'g't'n, D. C.				T	T			.32	.01										T	T		.01	T						T			0.34
Woods'k College					T			.24																								0.24
Norfolk, Va.				.16					.12																.22							0.62

NOTE.—"T" indicates a trace of rain or snow.

DAILY PRECIPITATION FOR NOVEMBER, 1892.

Stations.	1	2	3	4	5	6	7	8	9	10	11	12	13	14	15	16	17	18	19	20	21	22	23	24	25	26	27	28	29	30	31	Total
Baltimore		T	.02	.68				.46		.24	.49					1.19	.14	.54			T		.01	T				.24	.67			3.85
Barton Crk. Spr.		.26	.10	.36					1.40	1.26						2.34	.10	.06	.24				T		.50			.05	.10			5.30
Boettcherville						.10	.10				.10							.40	.10				.05					.06		.01		3.70
Cumberland (1	.10		.08				.10	.60	.08	.05					.04	.04	.80	.94	.18								.52	.02				3.58
Cumberland (2)	.08	.07	.07	.32					.13	1.41					.14	1.05	.83	.88	.14													2.18
Darlington				.48			.17		1.17	1.05				.80	2.60													.39				4.61
Denton				.18	.08			.02		1.53					.14	.24	2.07						T					.32	.13	.05		6.68
Dist. Reg. D. C.			.54		.61	T		.17		1.84					1.35	.63		.48	.08		.02		T					.33	.21			4.70
Dover, Del		T	.10	.27			.10		1.46	1.20					.08	.07		.62				T	T	T				.12	.07			5.79
Easton			.06	.08								T			.40	.65		.30			T							.15				4.10
Fallston			.06	.42			.40		.14	1.20					.08	.14		.87	.66		T	T	.02					.51	.05			6.36
Frederick					.03										.14	.70		.58					T					.04	.42			4.06
Great Falls								.12	1.25	T				2.50			.10						T					.35				4.11
Jewell		T	T	.40						2.05					1.15	.06	.87				T		.01					.47	.05			5.17
Leonardtown			.14				.04	.08	.04	.44		T					1.15						.04					.40	.42			5.31
McDonogh	.04			.53			.04	.04	.30	1.02					.39	.63	1.04	.04	T				.11					.65	T			3.12
Mt. St. Mary's	.03			.82			.25	.06	1.02	1.49	.03	T				.70		.16	.04				.11					.37	T			3.83
New Market	.04	.08	.12				.03	.15	.07	1.46					.10	.47		1.03					T					.24	.04			6.30
Rec. Res. D. C.	.10		.46			1.09		1.01	1.04						3.12			.00	.06	.31	T							.14	.08			3.66
Seaford, Del	.12	.01	.65				.40		.50			T			.70	.76		.66	.06		T		.06				.01	.11	.00			6.69
Solomon's															.40	.46	1.36											.41	.06			3.66
Taneytown	T	.25					.02	.10	.80	.48		T			.70	.06	.72	T	.04			.06						.41	.06			6.67
Wash'gt'n, D. C.	.17	.44	.17				.04	1.05	.46	.12					.82	.03	1.15	.28	.06				.01					.37	.09			4.08
Woods'k College			.58				T								.41	.03			.03				.01					.22	.02			2.96

Note.—"T" indicates a trace of rain or snow.

DAILY PRECIPITATION FOR DECEMBER, 1892.

Stations.	1	2	3	4	5	6	7	8	9	10	11	12	13	14	15	16	17	18	19	20	21	22	23	24	25	26	27	28	29	30	31	Total
Baltimore						.01		.07	T	.08			.82		T	T	.15	T	T	.67			T	T	.02	T					.02	2.28
Barron Crk. Spr						T		.06						.81			.83	T	T	.85		T	T	T	T	T					T	2.13
Boettcherville						.30				.08	T		T	1.40		.30	.20		.13	.10											.20	2.10
Cumberland (1)						.15	.15						.60	.40			.30		.23	.16											.30	1.64
Cumberland (2)						.12		.25				.06	.97		.08				T	.64					.06						.05	1.73
Darlington													1.30				.11		T	.53												2.23
Denton			.02										i.			.30	.06			.53												1.93
Dis. Res., D. C.						T		.04		.01			2.	1.45		.18	.15		.56	.54			T			.08						3.03
Dover, Del														1.23			.15		.56	.43												2.60
Easton						T		.20					1.			.18	.20		.56	.91		T	T	T	.04						.02	2.32
Fallston			T			T							1.	.40			.20			.40			T									2.21
Fenby						.05		.06					.60	.31			.20	.13		.36												1.60
Frederick								.10					.83	1.70	.06		.28	.13		.70												2.04
Great Falls														1.37			.36			.50												2.94
Jewell						T		T					.81				1.18			.52			T									2.25
Leonardtown								T					.60				.25			.48						.01						2.50
McDonogh			.08					.01					.97	.85			.16	.09		.36												1.37
Mt. St. Mary's			T		T	.02							1.75				.23			.59												2.00
New Market							.04	.08					.79	.47			.07			.64												2.67
Penn's Gr., N. J.			.03					.03		.16				1.77			.30			.60	.03				.04	.08					.05	2.13
Rec. Res., D. C.								T		.02			1.01				.30		.02	1.11			T									2.79
Seaford, Del			.03					T					.38	.42		.33	.14			1.13					.02						T	2.51
Solomon's						T		.52					.10	.99		.17	.10	T	T	T					.02						T	2.44
Taneytown													1.06	.70		.05	.23			.55					.07							1.77
Wash'gton, D. C						.02		.04					1.20				.05	T		1.04			T								T	2.82
Woodstock Col.								.06	T	.54				.04		.10	.05	T		1.27											T	2.32
Norfolk, Va																											1.75	.11				3.91

DAILY PRECIPITATION FOR JANUARY, 1893.

STATIONS.	1	2	3	4	5	6	7	8	9	10	11	12	13	14	15	16	17	18	19	20	21	22	23	24	25	26	27	28	29	30	31	To-tal
Baltimore	.91			T	.06	T			.14	.06	T	.46	.46	.02	.02	.02	T	T	T				T					T	.11		T	1.78
Barron Crk.Spr.	.72					.10			.07			.18		.15	.08				.10										.38	T		1.04
Boottoberville	T		T		.55	.80			T			.40		.15															.10	.10		1.20
Cambridge	.96				.20	.80	T		.80			.50		.10	.10				.15										.40		.12	.71
Cumberland														.10	.10																	.72
Denton	1.00			.10		.10			.20	.32	.30	.20	.08	.10	.04	.04			.10									.17		.06		1.90
Dist. Res., D.C.	.10	1.03			T	.10						.80			.07				.04				T						.32			2.16
Dover, Del.	.96				.12	.45			.30			.45																	.29			2.39
Easton	1.00																															1.29
Fallston	1.50								.30		.43	.21	.11		.21			T					T					.17				2.43
Fenby	.13			.20	.10	.10			.21		.20	.20																	.05			2.31
Frederick	.99			.33	.07	.07			.04	.33		.16		.04	.21				.03										.43			1.82
Glyndon	.14	.94			.13	.10			.40	.12		.60	.13	.16	.04				T										.11			1.19
Great Falls	.75			T	.14	.14			.18			.50		T	T								T						.30			1.84
Jewell	.95		.07	.13	.21				.18			.23		.08	.08				.08										.34			2.05
Leonardtown										.10		.36		.34	.25				.26										.15			2.18
McDonogh	.65				.13	.16			.10			.38	T	T	T				.25										.26			6.84
Millsboro, Del.	1.27				.30				.15		.83	.36	T	T	T				T										.34			2.18
Mt. St. Mary's	1.23					.22				.10		.37	.07	.33															.30			2.38
New Market	1.11	1.18			.11	.13			.08	.20		.35	.10	.09	.07				.02										.04			2.40
Penn's Gr., N.J.	.17			.20	.14	.10			.10	.20	.20		.10	.09	.06		.20	.40	.19	.20									.22	.06	T	2.23
Rec. Res., D.C.	.90				.08	.17	.07		.10		.37	.22	.30	T	.14				.20										.25	T		2.19
Salisbury	.67			.09	.08	.08			.40		.90	.90	.01	.10	.50				.10			T						T	.10	.10		2.21
Seaford, Del.	.50	.20	.10	.10	.50	.80			.02	.04		.02	.01	.10	.01														.34	.34	T	2.13
Solomon's	1.57			T					.17	.04		.44		.02					T										.06			1.58
Sunny Side	.90		.05	.12	.03	.04			.05			.20	.20	.02	.02														.30	.06		3.50
Tannytown	.85		.05	.03	.10	.20	T		.55		.50	.04		.10	.20				.08										.14	.14		1.96
Wash'gt'n, D.C.	.90		.03	.05	.10	.14																							.70	.11		1.85
Westminster	.13	.01	.09	.12	.12	.05		T																								1.73
Woods'kCollege																																2.97
Norfolk, Va.																																2.55

NOTE.—"T" indicates a trace of rain or snow.

DAILY PRECIPITATION FOR FEBRUARY, 1893.

Stations.	1	2	3	4	5	6	7	8	9	10	11	12	13	14	15	16	17	18	19	20	21	22	23	24	25	26	27	28	29	30	31	Total
Baltimore	.03	T	.08		T	.44	.08		T	.56	.28		1.21		.02	.09	.78	.09	T	T	.01	.21			.03	.08		.72				4.43
Barton Ck. Spr.			.40			.53	.10	1.23		.80	T	1.42				.33	T	.33	T	T	.50	.26			.01			.53				4.39
Boettcherville						.20			T	.06	.02	1.00	.01			1.40	.64	1.40										T				4.40
Cambridge	T					.05			T	.75	1.50	1.50	.01				1.00		.04		.40							.92				4.80
Cumberland (1)	.26	.06	.40			.19	.33		.61			.70	.70			.80					.50							.52				3.96
Cumberland (2)									.16	.80	.04	1.00	1.00			.10	.95	.10	.04		.44							.42				3.58
Darlington						.52	.23					.67	1.60			.75	.20	.75		T		.32						.95				4.71
Denton						.44	.06			.22	.40	.72	.80	.65		.10	.20	.10				.34			T			.63				4.38
Dist. Res., D. C.	.04		.01			.22	.81		.02	.22		1.53	1.53	.04		.81		.81		T	T							.50				3.53
Dover, Del.	.02				.10	.39				.85			1.86			.55	.58	.55			T	.34						.18				3.87
Easton	.05				T	.60	T		T	.50	.80	.21	1.34	.02		.88	.88	.88			.45	.45						1.10				5.43
Fallston			.05			.50	.40			.31	.13		.90			.58	.58	.58				.40						.90				4.24
Fenby	.04					.49			.02	.60	.19	.25	.67	.23		.14	.88	.14	T		.18	.23						.88				4.08
Frederick	.02	T	T		T	.23	.06		T	.34	.50	.66	.60			.78		.78	.04		.13	.20			.03			.74				4.51
Glyndon	T		T			.96	.20		T	.50	.50	.21	.60	1.34			.20				.45	.10						.62				3.78
Great Falls											1.24		1.02			.49	.49	.80				.16						.73				3.22
Jewell	.08		.06			.88				.53	.22	.22	1.16			.04	.30	.04			.66	.34			.05			1.02				4.60
Leonardtown		.06	.02			.43			.01	.42	.70	.70	1.01			.10	.66	.10			.13	.36		.20				.86				4.14
McDonogh	.12	.05	.01				.75			.38		T	.79			.75	.75	.80		T		.70						1.30				4.43
Millsboro, Del.													2.00									.86						.36				4.70
Mt. St. Mary's	.63		.06			.13	.16		.45	.12	.38	.08	.76	.04	.04	.73		.73			.32	.32			T			.36				4.01
New Market	T				.10	.52	.02		.01	.32	.96	.02	1.61		T	.71	.11	.71	T		.02	.35		.30	.01	.56		.78				3.47
Rec. Res. D. C.	.01	.01	.30			.59	.50		.20	.30	.69	T	.12	.67	.20	.50	.80	.50	.20		.20	.16	.30	T	.10			.40				4.85
Seaford, Del									.17	.62	.36	T	.80		.05	.01	.01	.07				.22						.34				2.60
Solomon's	.01		.61			.30			.01	.36	.41	T	1.48	.03		.66	.66	.07			.01	.32						.56				4.25
Sunny Side	.05		T			.30	.06			.90	.58	T	.90			.10	.65			T	.54				T		.96	2.25				4.30
Taneytown						.45	.06		T	.60	.90	.40	1.10			.49		T	.20													4.45
Wash'g't'n, D. C.	.01	.01				.12	.07		T	.11	.30	T	.07	.45	.01	T	1.50	T			.10	.06			T		T	.29				4.04

DAILY PRECIPITATION FOR MARCH, 1893.

Stations.	1	2	3	4	5	6	7	8	9	10	11	12	13	14	15	16	17	18	19	20	21	22	23	24	25	26	27	28	29	30	31	Total
Baltimore				.18			T		.53	.12	.24			.02	.06		.05	T		.01	.07	T	.06	.04								1.38
Barren Crk.Spr.			.20	.36					.74		.37				.60			T			T		.08									3.38
Boettcherville	.70		.20	.20	.33				.80	.23	.40		T				.60	1.00		T	T			.60		T						2.40
Cambridge			.18					.60	.70	.42			1.10								T		.42									4.14
Cumberland (1)			.17						.80									T					.04									1.28
Cumberland (2)				.05						.70			.15				.02		T		.10		.08									1.00
Darlington				.60					.80		.32		1.14		.70		.60	.16		T		.22	.13									2.10
Denton	.64		.01	.47	.03				.48		.32						.40	.25		.02	.07			.15								3.27
Dist. Res., D. C.				.50			.09		.81	.10	.53		1.16	.97					.02					.05								3.32
Dover, Del				.60					1.06	.36	.29		.10				.08		T	.15												3.34
Easton				.17					.55	.30	.20		.10	.20		.02		.01	T	.09												3.44
Fallston	.45			.45					.70	.27			.05	T		.04				.10			.03									1.55
Fenby			T	.14				T	1.07	.21	.82		.05		T		T		.08	.00		.00	.14	.06								1.85
Frederick				.36				.50	.59	.16			.75		T		.04	.10	T		T		.10	T	.03							1.84
Glyndon				.15				.78	.34	.20	.23		.85		.20		.60			.02	.00	.05	.09	T								1.67
Great Falls			.01	.13				.25	.58	.73	1.15		.09	.42			.24			.04	.06	.08	.04		.07							1.70
Jewell				.58					1.00	.09	.32		.30							.30	.20	.06	.18	.10								2.80
Leonardtown				.50				1.01	.30	T	.42		.10				.60															2.68
McDonogh	.86			.34					.45	.30	.24		T	.25		.17		T		.07	.07										2.24
Millsboro, Del			.04	.73				T	.73		.34			.82			.42			.05	.15											3.09
Mt. St. Mary's			.08	.65					.63	.17	.61		.01	.15		.12		.10	.01	.10		.08	.05	.65								2.42
New Market				.43	.14				.04	.01			.02				T				T		T	T								1.90
Rec. Ho, D. C.			.01	.47			T	T	.96	.06	.16		.05	.21			.12	.10	.01		.10		.08	.03								2.26
Seaford, Del				.40					.50	.30	.30			.12			.02	.01			.10			T								3.29
Solomon's				.12					.80	.33	.83						.02		.01		.05			.07								2.67
Sunny Side		.02	T	.53				T	.81	T	.27		T	.18			.22	.02			.03	.04	.27	.02	.38	.59						1.24
Taneytown																																1.99
Wash'g't'n, D. C.																																1.83
Westminster																																1.36
Woods'n College																																1.62
Norfolk, Va																																3.32

NOTE.—"T" indicates a trace of rain or snow.

DAILY PRECIPITATION FOR APRIL, 1893.

Stations	1	2	3	4	5	6	7	8	9	10	11	12	13	14	15	16	17	18	19	20	21	22	23	24	25	26	27	28	29	30	31	Total
Baltimore		T				T	.15	.01	.11	.31	.01	.43	T	.05	.34		.02		T	1.03	.11	T	T		.12	.01	.74		.54			3.62
Barron Ck. Spr							.25						1.87							.50	.80						.74		.96			3.42
Benedict								.20	T					1.08						.50	.20	.10			.24	.60	1.25	.20	.25	T		3.29
Boettcherville			T	T			.22	.08	.86	1.30	.20	.50	.10	.20	.20				.15	.50	.20	.04				.52	.12		.39	.15		4.10
Cambridge (2)				T						.13	.13	.12		1.40	1.40					.64	.10		T		.19				.18			3.47
Cumberland (2)							.20			.33	.35	.23	.23	.76	.83	.37				.81	.03				.15				.19			3.9x
Darlington							.16	.15	.16	.46	.25	.12	.25	.25	.37					.66	.03				.20	.14	.63		.03			3.18
Denton									.18	.08	.25		.13		.33			.07		.22	.50	.02			.20	.14	.35		.07	.08		3.77
Dist. Res., D. C.									.27		.30		.16	T	.59	.59					.62						.70		.12	T		2.85
Dover, Del	T						.23	.15	.30	.30	.02	.04		T	.73		T			.57		T			.08	.30	1.56		.06	T		3.59
Easton	T						.20		T	.90	.02	.40	.10	.82	.82					.87	.24	.02			.10	.16	1.36		.10	.10		3.38
Fallston							.10		.04	.30		.60	.10	1.00	.70	.70	T		T	.90	.50				.08	.16	.49		.13	.13		4.90
Fenby						T	.12	.12	.05	.43	.26	.45	.15	.45	1.54	.54	T		T 1.42	1.17	.11	.11			.44	.06	.63	.06	.04	T		4.80
Frederick							T	.17	.10	.38		.24	.24	.31	.31	.36	.25	.23	.23	.06	.54	.05			.25	.14	.63	.06	.08	.02		4.83
Glyndon	T														.50	.50	.12			.10					.25	1.00	1.00		.08			2.69
Great Falls									.10	.20		.15	.02	1.93	1.93					.13	T				.12		1.20					2.56
Jewell						T	.17	.58	.13	.43		.47	.08	.24	.43		.12		.17	.97	T	.11	.17		.12		.78					4.16
Leonardtown							.33	.06	.18	.08		.05	.01	1.34	1.34				1.06	1.06	.60	.60			.22	.03	1.35					3.72
McDonogh					.13		.16	.15	.15	.02	.27	.48		.16	.46				1.50	1.50	T				.11		.70					5.31
Millsboro, Del.	T						T	.56	.10	.38		.60			1.00	1.00			1.57	1.57					.57		.44	.06	.14			3.61
Mt. St. Mary's							.04	.10	.13	.20	.20	.36	.40	.40	.93	.03	.02	.03	T	1.36	.43	T	.17		.57	.47	.58	.08	1.03	.05		3.36
New Market							.19	.09	.22	.07		.14	.14	.33	.34	.34		.14	.04	.04	.56	.11			.15	.15	.29		.06	.06		6.75
Oakland	.28									1.17	.20	.36			1.63				.04	.41	T				.20		1.29		.36			2.68
Res. Res., D. C.				T		T	.09	.03	.03	.13		.03	T	T	1.46	.02	.62		.28	.28	T	T	.01		.43	.65	.84		.16			4.46
Seaford, Del				T		T	.10	.10	.19	.26	.27	.10	.10	.46	.40		.09		.97	.74	.17	T	.01		.13	.05	.30		.17			3.54
Solomon's		T		T			.27	.16	.15	.02		.02	.02	.46	.41				1.29	1.20	T	T			.14	.79	.70		.06			5.38
Taneytown							.14		.06	.38	.06	.07			.17				.02	.89	.01				T	.50	1.14		.60	.64		3.21
Wash., D. C				T	.05	.01	.14	.02	.10	.01		.01	.06	T	T	.02	.02	.04	.04	.16	.01	.06			T	.63	.23		.25	.84		3.81
Woodstock						.02	.10			.06	.06	.06		1.14	1.14				.25	.25						.38	1.08		.26			3.40
Millville, N. J.																																2.42
CapeCharles, Va																																2.07
Norfolk, Va																																2.99
Warsaw, Va																																

NOTE.—"T" indicates a trace of rain or snow.

DAILY PRECIPITATION FOR MAY, 1893.

Stations	1	2	3	4	5	6	7	8	9	10	11	12	13	14	15	16	17	18	19	20	21	22	23	24	25	26	27	28	29	30	31	To-tal.
Baltimore	.02		1.13	1.18	T								T	.24	.02	.31	T			T	.05	.13	.24	.10		T	.15	.04	.03			3.78
Barron Crk Spr		1.13		1.54			.25						.31	.56			T				.26			.67		.17	.08		.67			4.29
Benedict			1.60		.60								.81			.08			.60		.60	.18				.17			.70			4.14
Boettcherville	.10		1.50	.50	.20								T			.40	.70				.77		.02		.78		.25	.10	.34		T	4.99
Cambridge	.16		1.50	1.50		.64							.38			.45				.62	.85		.08		.73		.12	.24				4.11
Cumberland (1)	.07	1.40		.60										.01	T	.35									.73		.25		.34			4.50
Cumberland (2)		2.32			.72								.03			.47					.24		.48				.12		.60			4.37
Darlington															T	.62	T						.44	.17					.13			3.97
*Denton														.48		.11	.10				.24	.76					.12		.28			2.49
Dist. Res., D.C.		.04	2.33		.08								.12			.62	T						.41						.74			4.81
Dover, Del	.01		1.52										.31			.37				T	.33	.67		.45			.33	.32	.54		T	9.50
Easton			1.60		.04										.54	.45	T				T		.10				T	.10	.64			4.18
Fallston	T		3.45		.60								.02			T					.10	.67		.30				.10	.03			5.66
Fenby	T	.10	1.90	1.10	.90										2.30		T				.11		.10			T		.10	.08			8.60
Frederick	.13		2.61	.54									.37	.14		.52	T			T	.24		.19	.06			.08	.03				4.78
Glyndon	.02	2.00	1.50	.30									.28	.14		.39	T			.05	T		.19			.09	.11	.03	.11			5.08
Great Falls		1.12	1.03			.03									T	.17	.71				.37	.89		.13		T		.06	.76			3.44
Jewell		1.76		T									.55			.10					.78		.75				.35		.73			4.62
Leonardtown	.06	2.57		T	T								.21	.04	.08	.60				.93	.76	.59	.56	.45			.08	.02	.56			4.71
McDonogh	.06	1.25	1.25	.12	.30								.34	.04	T	.33	.02			.02	.30		.01			T	.03	.19	.54			4.57
Milford, Del		.03	1.00	.95	.09								.09	.21	.01	.58				1.06	.18	.15	.18	.15		T	.02	.11	.37			3.09
Millsboro, Del	.03	.32	.08	.55									.47	.01	.04	.93	T			.04	.43	.01	.78	.23			.07					3.64
Mt. St. Mary's	.32		2.10	.30		.90							.40			1.22	T				.20					.08		.07				4.24
New Market	.16		2.10		.60								T	.16	.14	.60	.35	.15		.93	.84	.56	.11			.12	.11	.03				5.30
Oakland	.09	1.08	.00	.50	.60								.04	.04		.21	.33			.11	.11	.40	.64				.06					4.78
Rec. Res., D.C.		.07	2.85										.04	.24		.55	T			.02	.44		.40	.21		.07	.11	.11	.14			4.77
Seaford, Del			1.70	T	T								.40	T	.01	.37	T			T	.28	.28	.15	.55		.01	.48	.24	.47			3.80
Solomon's	.09	1.05	.67	T									.13	.32	.01	.45	.53			1.06	.68		.01				T	.16	T			6.56
Sunny Side	.19	T	1.40	1.00	.20								.30	.11	.04	1.27	T			.04	.35	.15	.15				T	.08	.19			3.66
Taneytown	.21			.29	.31	.85	.16						.45			.17					.61		.10				.29	.05	.03	.44		4.44
Upper Marlboro	.12		2.00		.10		.07						.62	.02	.09	.33	T				.47	.16	.05	.10		T	.05	T	1.00	.40		5.15
Wash'gt'n, D.C.	T	T	1.76	1.02												1.30					.20	.06	.40				T			.40		4.40
Woodstock				2.60	T									.30										T			.29	.14	.40	.60		3.97
Birdsnest, Va		.60	.40		.60				.85				.01				.90						.10	.06								6.74
C. Charles, Va		.80	.48					T				.16			.77					.08			.40			T	.29		.48	.06		3.87
Norfolk, Va		.46	.75	.10	T		T	.12	T				.09	T	T	T	T				.22						74	1.38	1.30	.06		6.74
Warsaw, Va			1.84			.02	.34		.78	T			.01			.66								.20								4.37

° 1st to 9th inclusive, missing.

DAILY PRECIPITATION FOR JUNE, 1893.

STATIONS.	1	2	3	4	5	6	7	8	9	10	11	12	13	14	15	16	17	18	19	20	21	22	23	24	25	26	27	28	29	30	31	Total.
Baltimore........	.04	.62	.01	.06		.09					.01		T		T	.10	.03				T	.44	.20	T		.68			.05	T		2.26
Barron Ctr.Spr .		.81	T		.01		.02					T					.06				.11		.11		T	.11	.02		T			.72
Benedict.......				.50		.70										.20						.20					.30	.10	.10			1.30
Boettcherville..	.80		T			.10															.10	.40	.20			.30	.30	.10	.10	T		3.40
Cambridge......	1.64	.55		.17		.03															.30	.58	.20			.30	.52					2.94
Cumberland (1).			.29	.36	T									.04							.24	.12				.58	.65		.10			1.98
Cumberland (2).	.56		.12		T																					.31						2.12
Charlotte Hall .		.41	.01	T	.29	.62	.02	.03																								1.66
Darlington......		1.86		.44		.66		T				.02	.01	.04	.06		.10				T		.40	T	.29	.46	.30		T	.06		4.74
Denton..........		.65	.18		.08	.25											.10	.08				.18	.50	T	T	.25		T	.03	.28		1.80
Dist. Res. D. C.			.77	T		.08																.40	.42	.10	.4	.12						1.68
Dover, Del......	.67			T		.31						.04				T					.10	.09				.16						1.31
Easton..........	T	1.65	T	.09		.63											.04					.24	.58			1.00		T				1.48
Fallston........	1.90		.03	.10	.07	1.70	T				T					.16			T	T		.98				.80	.10	.06	.50			4.51
Fenby..........		T	.03	.18		.39	T								.08							.10	.16			.20	.10	.18	.40			5.60
Frederick.......	.04	.71	.18	.04	T	.04									.05	.18					T	.16				.30	.16		.34			1.44
Glyndon........		.98	T			.50	.28					.06			.08		.06	.12			T	.21	.06			.77	.06	T				2.38
Great Falls.....		.12	T	T	1.82	.60									.06	.18	1.28				.01	T	.18	.02		.65	.65		.03			1.85
Jewell..........																	.16				.35	T	T			.30	T	T				1.04
Kirkwood, Del..	.20	.49	.02	.16	.02	.07		T		.08		.02					.36		.02		.23	.05	.31			.62	T		.03	.06		2.00
Leonardtown...	T	.29	.12	.40	.10	1.68	.29				T						.15				.29	.66	.50		.02	.45	.03	.06	T			2.48
McDonogh......	1.32	.80			.02	.06					.26			T		T	1.28	.68			.71	.84	.15			.14	.03	.31		.40		3.60
Milford, Del....				T		.25																	.18			.30	.16	.18	T			1.73
Millsboro, Del..	.28	.07	.02	.16	.02	.07	.65	.06				.06					.16			.02	.01	.06		.02		.65	.08					1.68
Mt. St. Mary's..		T	.40		.02	1.80	1.16							T							.28	.66	.24		.06	.19	.31	.18	T	.60		.77
New Market.....	.31	.15			.07	.13					T						.14				.71	.84				.14				.40		2.57
Oakland........	.31			.03		.29	.29														.02	.01	.04			.23	.88	.02	.02			3.95
Rec. Dis. D. C.	T	1.72	.38		.02	.67	.01				.86			T	.05	.11	.22			.43	T	T			.33	.23	.10	.02			1.87
Seaford, Del....	T	.48		.02	.05	.06	T				T					.30	T			.05	T	.06	T		.06	.25	.31	.13			1.81
Solomon's......	.03	.66	T	.08		.16		.06			.26					.04	.03			T	.82	.82	.03			.68	.32	.02	.01			3.38
Sunny Side.....	T		T			.75	.65	.06						T		.20	4.80			.05	.06		T	.06		.63	T	.10	.36			3.96
Upper Marlboro		T	.02			.47	.97									.01	5.70			.10	.10		T		T	.40	T	T	T			6.78
Wash'n'n. D. C.	.30				.06	.29	1.16						.02				1.31					.10	T			.18	T	T	.86	.35		8.36
Wood's College	.78					.08	.08										.64						T				T		.01			1.09

NOTE.—"T" indicates a trace of rain or snow.

DAILY PRECIPITATION FOR JULY, 1893.

STATIONS.	1	2	3	4	5	6	7	8	9	10	11	12	13	14	15	16	17	18	19	20	21	22	23	24	25	26	27	28	29	30	31	To'l
Baltimore.		.01				T	.60	1.00	.01			.01		T	.00		T	.03								.21	T		T		.02	1.90
Barren Crk Spr	.51	.51					.61	.01				.10			.07										.24					.02	2.20
Boswell.	.00		T												.12			.00								.18	.21				.46	1.17
Beechetherville	.10	.40	.14	.31	T	.40						.10	.20	.10		.10											.40	1.50
Cambridge	.30	T			.18	.22	.40				.50	.20	.18		.10	.10									.16	.01		T	7.12
Charlotte Hall.	.45	.20			.28	.15	.69			.30	.30	.05	.24	.08	.10	.07	.01										.65	2.05
Cumberland (1)	.47	T	.10		.44		.01			.10	.10	.04	.07	.00	.07										.05			1.80
Cumberland (2)	.42				.21					.54	T	.16		T				.16	.01						.42				.74	1.40
Burlington		1.10			.36	.21	.29					.35																	.05	2.06
Boston		.10		.26														.05							.65			4.07
Hd. Res., D. C.		.15		.21		.12	.12	.40			.10	.62	.40	.21				.42	.08							1.47
Dover, Del				.10		.12	.00					T	.12					.53											6.29
Easton	.40	1.50	.81	.21	.09					.90	.08	.11	.88					.58	.10										.00	4.20
Fullston		.60		.12	.64			.80			T	.47		.00	.20	.20											.10	2.08
Fenby	.29	T	.28										1.40	T	.04															T	1.80
Frederick	.07		.62	.01	.64					.83	.16																				1.80
Great Falls													.44																		2.10
Jewell		.60										.88					.76						.17							.40
Leonardtown	.44				1.42	.86		.40	.24	1.93	.01	.14	1.28					.25					.17					.04	8.64
McDonogh	.67			.40					.57	T															.01			1.04
Milford, Del	.65	.30		.47	.36	.01	1.81	.70															.00	.00			2.19
Millsboro, Del	1.35			.47			1.94									.25	.04			.07						.17			2.47
Mt. St. Mary's				.48	.83											T											.30	3.68
New Market	.10	T		.63	T		.40						.03						T						.00			1.70
Oakland	.30	.17												.42					.15										T	2.00
Rev. Res., D. C.	.60		.80				.70	.01	T				.10						.07					.24						3.41
Seaford, Del	.05		.25	.43		.11	.42									.00					.00	T			.25				T	17.01
Sunny Side	.25	.25	.43	.11	.04	.14						.14												.10					.02	5.14
Solomon's	.60	.12	.71	.13	.82	.62											.00									.30			1.82
Upper Marlboro	.63		.18	.42			.01									.04	.12									.43			4.04
Washington, D.C.	.60		.33	.14													T								.07			.44
C. Charles, Va	.20	.30	.18		T																						.14			6.11
Norfolk, Va	.40			.05			.48	.05																	.07			.05
Warsaw, Va	1.70	.70			.95														.40							.22			5.01

NOTE.—"T" indicates a trace of rain or snow.

DAILY PRECIPITATION FOR AUGUST, 1893.

STATIONS.	1	2	3	4	5	6	7	8	9	10	11	12	13	14	15	16	17	18	19	20	21	22	23	24	25	26	27	28	29	30	31	To'tal
Baltimore				.08	T	.04						.05					.03		T	T			T	.61			T	T	1.02			1.81
Barron Ck. Spr.	.55			.05		.06											.05						.02	1.12				.50	.85			2.67
Benedict			.32									.05					T			.10	.10		.78									3.16
Boettcherville			.10							.10	.10						.50									.30		.10	3.00			4.30
Cambridge					.05							.68					.15						1.63					.10	.30			1.66
Cumberland (1)			.04									.10					.56		.66				.17				.36	.36				4.03
Cumberland (2)						.67		.41				.08					.32										.60	.65	2.65				3.74
Darlington				T								.24					.08			.39		1.33		.18				.71				3.60
Dist. Res. D. C.					.16	.63						.25					.33	.12		.10	.08		.36						1.00			1.85
Dover, Del					.28	.02						.07					.23	.06					2.05						.70			3.08
Easton				1.57													.63			1.08			.20	1.36					.24			4.24
Fallston				.30		.35											T															6.26
Fenby																	.10		T										.40			4.30
Glyndon			.13	.06													.02									.01						2.68
Great Falls					.16										T			.07	T		.08		.44						2.10			2.81
McDonogh				.15	.05	.05		.47									.38	.01		.57				.03	1.98			.56	.99			3.82
Milford, Del			.14									.40						.07		1.45	T		.11	.80	.04			3.65	.69			3.61
Millsboro, Del						.10						.18					.15			.28			.59	.36			T	T	1.09			4.00
New Market				T											T		T			.06	T		.22	T				.12	1.21			2.00
Oakland	.04											.50					.27		T	.14		T	.19			.01		2.06				3.71
Rec. Res., D. C.				.08	.14		.78										.10	.07			.08		.08						.79		T	2.86
Seaford, Del				.04	.04							T							T		T		.99	.51				.86	1.76		.01	2.64
Solomon's	.10		.14			.07						T					.12	.01		1.47			.11	.80		.04		.08	.11	.17		3.07
Sunny Side			.06		.04							.44					.72			.28			.59	.36				3.65	.62			3.68
Upper Marlbo.			.14	.09	1.15							.40		T			.21	.07					T	T				T	T		T	3.40
Wash., D. C.	T		.15	.02	.21							.18	T				.15		T	.00	T		.22	T				.12	1.21			2.32
Woodstock			.10															.14										2.06			T	2.88
Bielhurst, Va.	.40			T	.56							.56					1.40			.76		2.00									.50	4.65
C. Charles, Va.												1.34					.24			.14			.90					.75			.61	5.25
Norfolk, Va.	.46		.35	.97								.10					.02	.19		1.46			T	.04				.02	.72		.02	35.71
Warsaw, Va.	.12		.22									1.55								1.55			.08						.64			2.93

NOTE.—"T" indicates a trace of rain or snow.

DAILY PRECIPITATION FOR SEPTEMBER, 1893.

STATIONS.	1	2	3	4	5	6	7	8	9	10	11	12	13	14	15	16	17	18	19	20	21	22	23	24	25	26	27	28	29	30	Total
Baltimore	.24	T				.02					.20	T	.21	.54	.29	T						.12		.28	.01						1.80
Barron Crk. Spr.	.38									T	.13		.15	.38	.66									.10		1.66					3.01
Benedict	.47					.20	.40				.30		.33									.19		.40	.48						1.41
Boettcherville	.40					.10	.30	.28		.06	.10	.20	.29	.30	1.00	1.00			T			.07		.17	.30	.50			.39		3.10
Cambridge										.00	.10	.50	.30	.30	.43	.60						.08		.07	.07						2.94
Cumberland (1)											.18		.12	.12										1.04	.22					.07	1.99
Cumberland (2)											.17	.36	.01	.07	.56		.22														1.97
Darlington	.53													.63	.63	.01					.07	.04		.69	.04						2.94
Dist. Res., D. C.	.40	.92			.10	.10	.30							.26												T					3.96
Dover, Del.		.40																							.60	T					2.10
Easton	.46												.68	.10	.60	.30			T		T			.40	.69	.27					2.84
Fallston	.25					.02					.21		.20	.00	.60	.10						.10		.20		.34					9.70
Fenby	.20										.44	T	.33	.84		.30									.52	.34					2.00
Frederick	.10					.13				.10	.03		.85		.15	.23			T		.08	.10									2.64
Glyndon	.10	.20				.08		.05					.20	.22		.37								.48		.11					2.27
Great Falls											.10	.22		.54	.44	.00						.11		.50							2.01
McDonogh						.00								.43	1.31										.55	.54					4.32
Milford, Del.	.40													1.66	1.08									.70	.23	1.16					6.17
Millsboro, Del.	.44	.15										.04	1.12	1.40	.30							.12	.07	.01	.01						9.59
Mt. St. Mary's							.05						.30	1.10	.04	.54		T			T	T		.64							1.44
New Market	.25					.44				.25		.08	.30										1.00								1.44
Oakland		.06													.02	.03		.18					T	1.00	.10					.05	3.14
Rec. Res., D. C.	.04	.45								T	.21	.15	.07	.44	.02	.03	.18							.82		.28					3.76
Seaford, Del.	.40									T	.25		.25	.10	.05							.06		.54	.54	.70					2.56
Solomon's	.20										.44	T	.18	.04	.07	T			T			.10	T	.60	.60	.31					1.80
Sunny Side		T					T			.31	.12	.10	T	.37	T	T						.05	T	.42	.37	T			.11		2.06
Upper Marlboro'	.72									.16	.44	T	.49	.21	.04	.24			T				T	.92	.30						4.42
Wash'gt'n, D. C.	.00								T					.62	.00								T	.47							3.03
Woode'k College	.32								T			.31	1.25										T	.19							3.65
Birdsnest, Va.	.40	.20								.64	.49	.06	.12	.16	1.15							.40		.56	.44			.20			5.46
C. Charles, Va.	1.02									.85	.06	.10	.12	1.40								.01	.15	.11	T	T		T			8.27
Norfolk, Va.	.92								T	2.86	.15		.04	T		.80						.41		.65	.61	T	1.24				4.48
Warsaw, Va.		.37										.30			1.31										.43						

DAILY PRECIPITATION FOR OCTOBER, 1893.

STATIONS.	1	2	3	4	5	6	7	8	9	10	11	12	13	14	15	16	17	18	19	20	21	22	23	24	25	26	27	28	29	30	31	Total
Bachman's V'y				T		T						T	† 3.45									+					T					4.45
Baltimore				.95		T	.08						1.00	.60							.10?	.25	1.01			.06	.05	.01				3.44
Barron Crk.Spr.			+	+	.26	+	.34					+		.35								.20	1.06			+	.67					2.85
Benedict				.45																	.51		.30			.54		T				3.09
Bootchorville				1.70									1.50		.10							3.00				.93	.50					4.70
Cambridge				.30		.25							.11	.11								3.02				.61						3.61
Charlotte Hall				.41									1.86									.14				.52						4.48
Cumberland (2)			1.31	.15?									1.23	.11								.90				.37						4.37
Darlington			.15										1.08							.65			.89	.06		+	1.54					3.59
Dist. Res., D. C.				.04	.10	.06								1.08							+	+	2.09			+	1.17					3.30
Dover, Del				+	.06	+							.54									+	1.53			.23	T					4.77
Easton				.33		.06	.18						.93										1.93			.20	.10					4.04
Fallston				.21		.35							2.58									T	1.05			.17	T					5.15
Feaby				.50	.59	.10	.10						3.60										.70									5.96
Glyndon				.55		.02							2.14	T								.36	.80									3.93
Great Falls				.20									2.60																			4.29
McDonogh				.12		.06							3.53									+	1.77			+	1.36					3.88
Milford, Del.				.13		+	.14						.56								.21	.54	.85			.03	.43					3.87
Millsboro, Del.				.83		.03	.03						.60 2.13								.11	.13	.95			.16	T			T		2.68
Mt. St. Mary's				+	.66								2.00									+	.60			+	+					4.43
New Market				1.25	.52	.08							1.47						.97		.04	.04				.19	.21	.05					3.20
Oakland .01			+	.07		.07							2.54									+	1.09				.14					5.00
Rec. Res., D. C.				.20		.02							.54								T	T	1.16			1.18		T				4.36
Seaford, Del				.41									.91								.01 1.71	.44				1.38						3.12
Solomon's				+									1.50	.48						+	.90	T	.33	.09		+	.05	T				4.98
Sunny Side				1.10			.10						1.89									.82	1.04			.47	.15					5.02
Taneytown				.60		T							1.25 1.40										1.03			.86	.12					3.25
Upper Marlboro				.46		T							.37	.43							T		1.00				1.32					5.45
Valley Lee				.41		T							1.82	.45							T	.15	.95			.03	.10	T				6.40
Wash'gt'n, D. C.				.71		T	T							3.60								.40	.70			.40						4.11
Woods'k College				.50									.10 1.10								T 2.30				.75						5.60
Birdsnest, Va.				.20	.06	.30								.82								1.78				.40	.75					3.30
C. Charles, Va.				.82	.06	T	T																				.44					3.22
Norfolk, Va. .08				.06	.06	T							.90								.03 1.48		.86				.22			T		2.86
Warsaw, Va.				.30		.60							.93										1.53				1.17					4.04

NOTE.—"T" indicates a trace of rain or melted snow.   † Dates on which rain fell, but not measured.   * Rain gauge blown down.

DAILY PRECIPITATION FOR NOVEMBER, 1893.

STATIONS.	1	2	3	4	5	6	7	8	9	10	11	12	13	14	15	16	17	18	19	20	21	22	23	24	25	26	27	29	30	31	Total
Sunny Side			.18	1.70			T						T	T	.07		T	.09			.23	.11	T	.08			.06	.18†	.26		2.97
Oakland			.15	1.72	.26								T	T	.12		T	.00	T		.25	.08		.20			.25	.03	.19		3.44
Boettcherville			T	1.69											T				T		.10			T				.80	.20		2.90
Cumberland (1)				1.00	.70									.10							.30										2.80
Cumberland (2)			.46	.82				.33													.68						.53	.60	.10†		2.01
Mt. St. Mary's				1.25	.46	.12		.46						.50							.88						1.07		.12		4.67
Frederick				.91	.47				1.40				.66	.11					T	1.07	.89							.73			3.06
New Market				1.54			T						T	.45					T		.86						.86				2.7
Taneytown				1.40			+	.40					.10			.30					+	.60					+	.70			3.40
Bachman's V'y				2.33	.30			.30						.65	T						1.10						1.66	1.10			4.03
Fenby				1.00					.40										T	.04	.40	.53									3.30
McDonogh			T						.30				1.43	.82					T				T				.80	.70			2.69
Woodstock Col			T	.20	.35			.29					.04	.66	T		T		T		.35	.11					.85				3.70
Baltimore			.02	.14	.21	T		.37	1.08				.03	.03	.58		T				.68						.67				3.76
Fallston			T	.90	T			1.32					.73	.73			T				.82										4.53
Darlington				.68				1.07	1.40					.06	.39		T				.44										3.87
Great Falls					.37	.12			2.10				T	.05	.34		T				.35	.44		T			.90				3.58
Dist. Res., D. C.					.43	.05			1.93					.64	.56		T				.30	.35					.52				4.00
Rec. Res., D. C.					.90			1.84	1.05				T	.05	.60		T				.43	.44					.65	.93			4.20
Wash., D. C.			.10	.23				1.12	1.26						.67		.01			.04	.30	.02						.39			4.30
Upper Marlb'o			.05	.31		+		1.00					T		.76		T		T	+	.60	.36					.36	.49			4.42
Benedict				.18										.74																	2.86
Valley Lee				+	+	.02		2.00	.96						.72																3.27
Solomon's					+	.08								.15	.63		T				.63	.60	T	1.00			.31	.34			2.40
Chestertown			.18												.70		T		T		.30							.62			4.66
Cambridge									2.25					1.02	+		T		T		.63		1.00				1.00				6.25
Denton								1.68							.48		T			+	1.10	.48					.63	.47			2.35
Barren Ck. Spr.			.64						1.47						.80				.02		.57	.87					.01	.53			3.34
Dover, Del				+					.93						.74		T			+	.22						+				2.90
Milford, Del					.24				1.50						.60				T		.36	.48							.10		3.08
Seaford, Del					.07			.73	1.02						.37						.28	.14		T			.62	.54			3.15
Milsboro, Del					.18			1.02					T	.10	.74				T	+	.78	.35		.15			.60				3.41
Birdsnest, Va					T										.30						.15		.15				.84				1.90
C. Charles, Va					.08	+	1.20	5.02					T	.30			T				.25	.05					.84	.71			8.24
Norfolk, Va					.06	T	5.03	6.02	.72				.23	.01	.06						.17			.31							6.73
Warsaw, Va					.04		2.88														.61						.34	.39			3.42

NOTE.—"T" indicates a trace of rain or melted snow. † Dates on which rain fell, but not measured. * Rainfall not measured.

REPORT OF THE TREASURER.

Payments, April 1, 1892, to March 31, 1893.

Drawing base map	$25.00
500 lithographic copies of map	15.00
Drawing materials	1.75
Drawing isotherms and isobars on map	7.00
Printing weekly crop bulletin	300.15
Printing monthly report	586.81
Traveling expenses	5.60
Postage	1.19
Clerical help, mailing reports	6.50
	$949.00

April 1, 1893, to December 31, 1893.

Printing weekly crop bulletin	$320.00
Printing monthly report	668.10
Typewriting	1.50
Drawing isotherms and isobars on map	9.90
Clerical help, mailing reports	4.25
Postage to U. S. consuls and others in foreign countries	41.13
100 bound sets of climatic charts	243.00
Framing charts for World's Fair exhibit	15.55
2000 sets small charts for Biennial Report	135.00
Printing Biennial Report	495.74
Express	4.50
Traveling expenses	1.70
Miscellaneous expenses	1.81
	$1942.18
Estimated expenses from January 1 to March 31, 1894	$57.82

The sum of $2000 per annum was appropriated by the Legislature for the support of the State Weather Service. In the Director's report the reason is fully explained why only one half of this appropriation was spent in the first year. The work was much limited by the comparatively small amount of money available for printing. During the second year it has been possible to very materially increase the size of the publications relating to the climatic features of the State, and during the six months of the Columbian Exposition in Chicago the edition was much enlarged and five or six issues of the monthly report were widely distributed throughout the State, to the visitors in Chicago, and through the U. S. consuls to a number of public libraries and individuals in foreign countries. For the domestic distribution the reports went out under the frank of the U. S. Weather Bureau, but for the foreign distribution postage had to be paid as on ordinary mail matter of that class. That this distribution was appreciated is shown in a large number of letters which have been received.

As this report has to be published in December for presentation to the Legislature, it is impossible to give a full financial statement, as there remain three months for the work to be carried on under the appropriation. It has been necessary, therefore, to estimate the expenses of the current monthly reports for the three months from January 1 to March 31, 1894.

Duplicate vouchers of all payments which have been made are on file in the office of the Comptroller of the Treasury at Annapolis.

MILTON WHITNEY,
Treasurer.

INDEX.

.

www.ingramcontent.com/pod-product-compliance
Lightning Source LLC
Chambersburg PA
CBHW021813190326
41518CB00007B/569